Damselflies of Alberta
Flying Neon Toothpicks in the Grass

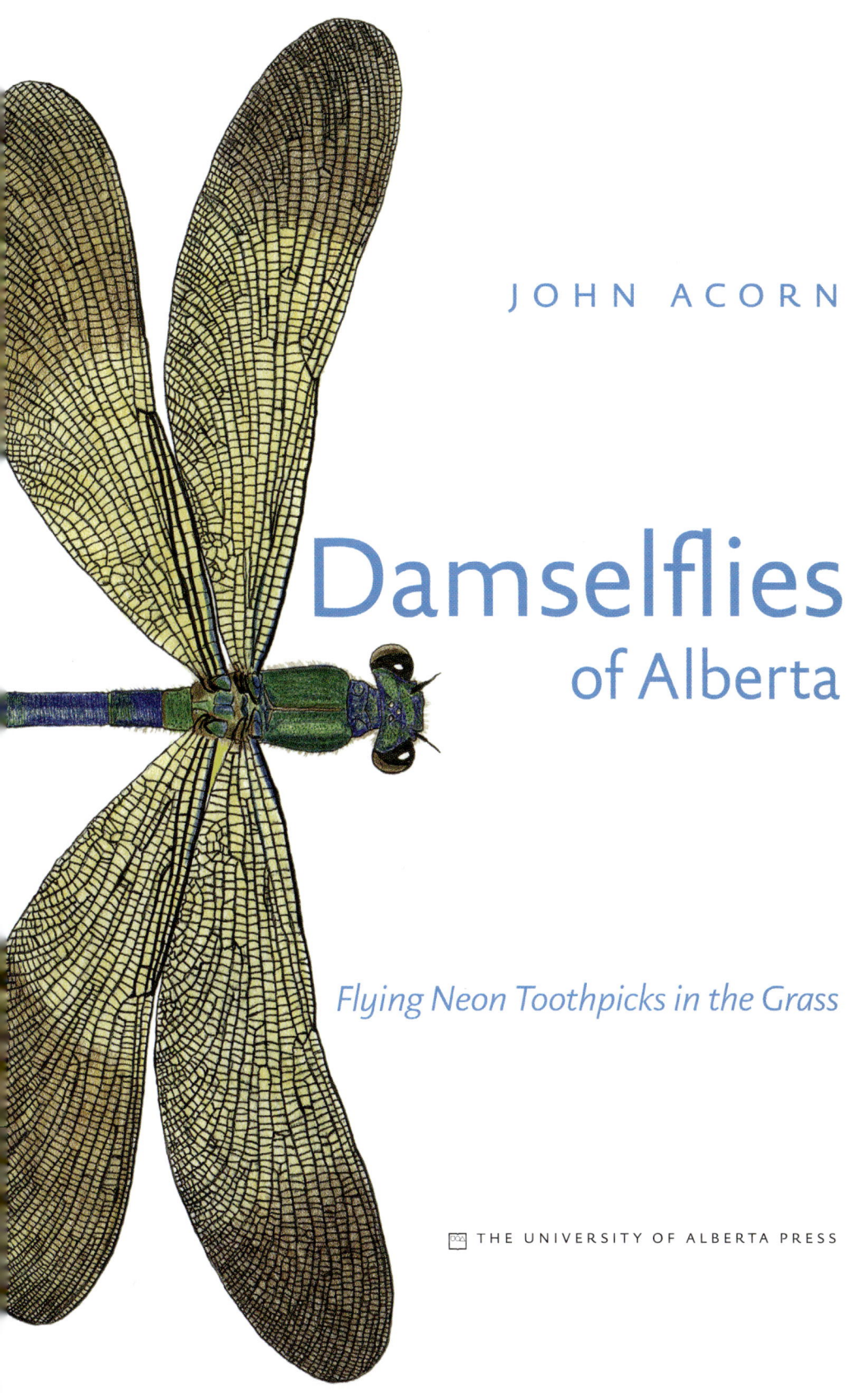

JOHN ACORN

Damselflies of Alberta

Flying Neon Toothpicks in the Grass

THE UNIVERSITY OF ALBERTA PRESS

Published by
The University of Alberta Press
Ring House 2
Edmonton, Alberta T6G 2E1

Copyright © John Acorn 2004

LIBRARY AND ARCHIVES
CANADA CATALOGING IN
PUBLICATION

Acorn, John, 1958–
    Damselflies of Alberta : flying neon toothpicks in the grass / John Acorn.

(Alberta insects series)
ISBN 0-88864-419-1

    1. Damselflies—Alberta. I. Title.

QL520.2.A4A36 2004   595.7'33
C2004-901540-0

First edition, first printing 2004
All rights reserved
Printed and bound in Canada by Kromar
    Printing, Winnipeg, Manitoba
∞ Printed on acid-free paper.

No part of this publication may be produced, stored in a retrieval system, or transmitted in any forms or by any means, electronic, mechanical, photocopying, recording, or otherwise, without the prior written consent of the copyright owner or a licence from The Canadian Copyright Licensing Agency (Access Copyright). For an Access Copyright license, visit www.accesscopyright.ca or call toll free: 1-800-893-5777.

The University of Alberta Press gratefully acknowledges the support received for its publishing program from The Canada Council for the Arts. The University of Alberta Press also gratefully acknowledges the financial support of the Government of Canada through the Book Publishing Industry Development Program (BPIDP) and from the Alberta Foundation for the Arts for its publishing activities.

All photographs by John Acorn, with the exception of the following:

Carroll Perkins: pages 42, 63, 69, 71, 72, 74, 75, 78, 117, 125
Jonathan Hornung and Christine Rice: pages 16 (bottom left), 30 (bottom), 45 (top)
Red Deer Archives: pages 22, 25 (top)
Sid Dunkle: pages 52, 57
Blair Nikula: page 54

To Gordon Pritchard, and his many fine students.

Contents

Preface ix

Acknowledgements xi

1 | Flying Neon Toothpicks? An Introduction to Damselflies 1

2 | A Day in the Life of a Damselfly 9

3 | Damselflies and Wetlands 19

4 | The History of Damselfly Study in Alberta 23

5 | How to Study Damselflies 35

6 | Damselfly Conservation in Alberta 43

7 | The Damselflies of Alberta 51

Appendix 1: Checklist of Alberta Damselflies 129

Appendix 2: Key to the Adult Damselflies of Alberta 131

Appendix 3: Helpful Sources for Damselfly Study 139

Glossary 141

References 146

A Gallery of Damselflies 151

Preface

This is the second in what I plan to be a series of books featuring selected groups of Alberta insects. The first book of the Alberta Insects Series was about tiger beetles, and the volumes that follow will, if all goes well, treat ladybugs, the larger moths (what I call the "big snazzy moths"), the dragonflies proper, and possibly the grasshoppers as well. *Tiger Beetles of Alberta: Killers on the Clay, Stalkers on the Sand* was an experiment of sorts, and in most respects I think it succeeded. One tiger beetle specialist called it "a splendid mix of science and élan." Another entomologist wrote to me to say " WOW, is the tiger beetle book ever great! I have been standing in the hall like an idiot exclaiming to everyone who goes by about it." That made me feel good.

I think that most entomologists would love to write about their favourite bugs in a conversational fashion, if only the world of scientific publishing would let them. I'm very fortunate to be able to write in a less constrained style in these books. By contrast, in the "primary literature" of journal papers and technical works, all emotive language must be purged from the text. The result, predictably, is dull writing in which all authors sound pretty much alike. After more than two centuries of this sort of thing, scientific writing has become almost inaccessible to average readers, regardless of their intelligence or overall education. This saddens me, and I think it also weakens science by making its details seem less and less relevant to the rest of the thinking world.

Scientists, however, behave one way in print and another way in person, and in so doing they make it clear that they are no more Vulcan-like than anyone else. (I assume that most people reading this will be familiar with the Vulcans of *Star Trek* fame, and their so-called logical outlook on life.) In person, scientists are people like any others. They discuss scientific matters with little or none of the stylistic constraints that they show on the printed page. This leads many people to think that the conventions of scientific writing are largely trappings. The trappings of dull-style science give the ritual of science (if I may be so bold as to call it such) an aura of objectivity, and that is what counts on a purely subjective level—the aura.

It is my view that true objectivity (and objectivity is the usual justification for dry prose in science) is a matter of thinking clearly and self-honestly about evidence, and about the reasoning process that you use to link evidence to your conclusions. What the rest of your mind does at the same time is of no scientific consequence. The important thing, aside from correct logic and careful measurement, is self-honesty. In other words, you have to allow your deep curiosity about the nature of whatever it is that you study to continually push you to examine and reexamine your beliefs and conclusions. Without this sort of self-honesty, science becomes little more than a game of writing papers and applying for grant money. To me, the best way to demonstrate self-honesty is to write in a frank and personal fashion, admitting to one's influences, desires, uncertainties, and dislikes. I hope you agree. Fortunately, the study of damselflies has seen many such self-honest scientists, and I have been equally fortunate to fall under their influence.

Having expressed my views on this rather abstract subject, I would now like to invite you back down to earth, perhaps among the reeds at the side of a prairie pothole in mid-June. So join me if you will, and let me introduce you not only to the damselflies of Alberta, but to my own spin on their fascinating lives, and the community of real-life, subjectively biased people who have loved and studied them.

Acknowledgments

As with any book, this one was not a lonely project by any means. Over the years I have enjoyed many good odonatological discussions and times in the field with Miriam Abbas-Nejad, Kirsten Beirinckx, Randy Dzenikiw, Chris Fisher, Lee Foote, Ed Fuller, Su-Ling Goh, Hans Van Gossum, Jon Hornung, Doug McCauley, Layla Neufeld, Natasha Page, Christine Rice, Celine Sirois, Andreas Stange, Godo Stoyke, Lindsay Wickson, and, of course, my family: Dena, Jesse, and Benjamin. Rob Baker, Carl Cook, Lynda Corkum, Nicky Koper, Fred Korbut, Wayne Roberts, and Felix Sperling shared their notes, specimens, and observations. For inspiration and mentorship, I couldn't ask for better role models than Rob Cannings, Dick Cannings, Philip Corbet, and Dennis Paulson. I should also acknowledge my American colleagues and field buddies, Bob Behrstock, Blair Nikula, and Jackie Sones, all of whom helped motivate this book, albeit along streams and pondsides in Texas and Mexico.

For help with the historical portions of this book I acknowledge Charley Bird, Kelvin Conrad, Philip Corbet, Michael Dawe, Cedric Gillott, Gordon Pritchard, and Mark Wilson. And finally, I would like to tip my hat in appreciation to all of those who participated in the "Hyperboreal Odonatists' Guild and Social Club," especially Barb and Jim Beck, Charity Briere, Sean Bromilow, Chris Fisher, Kamal Ghandi, Kevin Hannah, Jon Hornung, Robert B. Hughes, Al Lindoe, Natasha Page, Greg Pohl, Christine Rice, Carole Patterson, Carroll Perkins, Sandi Robertson, Cindy Sheppard, and Terry Thormin.

These people have given me hope that a book about damselflies might not be such a weird idea after all. The University of Alberta Press deserves great praise as well, and I am happy to be working with such fine individuals as Alethea Adair, Alan Brownoff, Cathie Crooks, Jill Fallis, Michael Luski, Linda Cameron, and Yoko Sekiya. To Linda, especially, I owe a great thanks.

A lovely male taiga bluet, resting on ribbon grass, the way I first encountered them at age 5.

1

Flying Neon Toothpicks?
AN INTRODUCTION TO DAMSELFLIES

I expect that everyone has seen a damselfly at least once. Near ponds and the likes, damselflies often appear to be everywhere, especially among lush grasses. The most common ones are bright blue, and about the size and shape of a toothpick, with wings. From June through September, almost any sunny location in the province is likely to produce at least a few damselflies on a warm day.

My first memories of damselflies date back to when I was in elementary school, when I started noticing them along the back alley and in the back corner of my parents' garden, where the ribbon grass grew. I especially remember one sunny afternoon, walking to the bus stop to meet my grandfather and stopping along the way to admire what was probably a male taiga bluet resting on a white fence. At the time, it looked like it had every colour of the rainbow somewhere on its body, and it was with great surprise that I eventually discovered that these damselflies are merely blue and green.

I experienced my first spectacular emergence of American bluets at Gull Lake, the one and only time that our family rented a boat for an afternoon outing. Brilliant blue damselflies were skimming around everywhere I looked, just over the water's surface, and to be honest I remember that and not much else about the day.

Damselflies are related to dragonflies, and together they form the insect order Odonata (making those who study them "odonatologists"). Remember that insects form a class (Insecta, or Hexapoda) and that classes are divided into orders. In turn, the order Odonata is divided into three suborders. The first, Anisozygoptera, contains only two species and is found only in Asia. The second comprises the dragonflies proper—the suborder Anisoptera. The third suborder, Zygoptera, includes the damselflies.

In Europe, the entire order Odonata is referred to in English as "dragonflies," but here in North America we call the Anisoptera dragonflies and the Zygoptera damselflies. The order as a whole has no English name equivalent in North America, but most of us call them "odonates" or "odes" for short.

Odonates are some of the most primitive of all flying insects. Together with mayflies (the order Ephemeroptera) they possess wings that cannot be folded back flat over the insect's abdomen—a condition generally presumed to represent the state of affairs among the earliest flying insects, long before the days of the dinosaurs. In fact, the largest flying insect that ever lived was a primitive sort of odonate (*Meganeura monyi*—which was not a dragonfly in the modern sense) with a wingspan of about 75 centimetres.

Odonate and mayfly wings are attached directly to the flight muscles that propel them, as opposed to the situation in other flying insects in which the flight muscles change the shape of the thorax, and the wings flap as a result. The odonate condition is called paleopterous (ancient-winged), while indirect flight muscles are called neopterous (modern-winged). Paleopterous insects cannot fold their wings flat over their backs, while most neopterous insects can. Butterflies are a familiar exception to this rule, since they are neopterous but cannot fold their wings flat over their backs. This condition was acquired "secondarily" in butterflies, which evolved from moth ancestors with folded wings.

Odonates are predatory creatures, and this, along with their obvious structural differences, sets them apart from mayflies. As adults, odonates have four well-developed flying wings and an elongate body. On the head, they have huge compound eyes made up of thousands of individual visual receptors, each with its own lens.

As larvae, they are generally aquatic (although some elsewhere live in moist mossy places) and they are always predatory. Odonate larvae have the most amazing lower lips (*labia* is the plural, *labium* is the singular) in the entire insect world. The lower lip of an odonate larva is folded beneath the head. When the larva wants to capture another small creature for food, the labium shoots out to almost half the larva's body length and grasping jaws at its tip seize the prey. To my knowledge, the only other insects that possess this sort of labium are tiny rove beetles in the genus *Stenus*—often found along the shores of the very ponds and lakes that make up our finest damselfly habitats.

The larvae catch and eat a variety of small aquatic creatures (including extra-small fishes), and in general there are two types of larval feeding patterns. Some roam around on underwater vegetation in search of their prey, while others sit quietly and cryptically in wait. Generally, larvae that live in streams or under the threat of being eaten by a fish do the latter, while those in still waters without fish predators do the former. As well, hungrier larvae forage more, while those with lots of food around them are more content to wait for the prey to come to them.

By the way, the word *larva* (plural: *larvae*) is now the accepted term for any sort of immature insect. Some people prefer, however, to call the larvae of odonates "nymphs" (to indicate that they do not pass through a resting pupal stage in their life cycle) or "naiads" (to indicate that they are aquatic). I support

Top: Dragonflies, unlike damselflies, have semi-spherical heads, and quite stocky bodies. Bottom: To recognize a damselfly as such, look for the hammer-head, and the long, very thin abdomen.

This is a dragonfly larva, not a damselfly, since the three projections on the end of its abdomen are short and pointy, rather than long and leaf-like. Compare with the damselfly larvae on page 16.

the general use of the word *larva*, since it is one more way to reduce entomological jargon. Most damselfly larvae shed their skins about a dozen times before emerging as adults, and the number of moults can vary, even within a species.

To tell the difference between a damselfly and a dragonfly, look first at the head. If the insect is hammer-headed, with the eyes on either side of a wide head, then it is a damselfly. Dragonflies generally possess more bulbous heads with eyes that meet or nearly meet at the top of the head. Then look at the wings. If the hind wings are slightly broader than the front wings at the base (next to where they meet the body) and all the wings are held out to the sides at a right angle to the body, like the wings of an airplane, it is a dragonfly. If all four wings are of almost identical shape and are held vertically together over the back, or out to the sides at an acute angle to the body, then it is a damselfly. Once you get a feel for the general look of the two groups, these features will no longer be necessary for a quick identification, but when you are first learning to distinguish between damselflies and dragonflies, these are the features that work best.

You should also know that there is a group of vaguely damselfly-like insects found in dry, sandy habitats called ant lions (order Neuroptera, family Myrmeleontidae). To tell the difference between an ant lion and a damselfly, look for the long antennae of the ant lion—as opposed to the almost inconsequential antennae of the damselfly—and the dull brown colours of the ant lion as well.

As larvae, damselflies are also easy to recognize. Apart from the folding lower lip, they also possess three leaf-like gills on the tip of their abdomen. These gills can also be used for swimming, and a swimming damselfly larva wiggles its body from side to side. The larvae also breathe by flushing water in and out of the rectum, and as a consequence they literally breathe with their butts. For dragonflies proper, this is the primary means of respiration, but damselflies use both the rectum and the abdominal gills. Both, by the way (and I don't blame anyone for wondering about this), secrete a membrane around their feces so as not to foul their breathing apparatus—like having a built-in plastic bag dispenser when you take your dog for a walk. And damselfly larvae can develop other proctological difficulties as well. American bluets, forktails, and spreadwings are all known to have their back ends invaded by flagellate protozoa in the winter, only to lose these freeloaders with the first skin shedding of the spring.

The damselflies of Alberta belong to three different families. In fact, the three families of damselflies in our fauna represent the three main branches of the evolutionary tree of damselflies in general. Even though we only have 22 species here in Alberta, we still have representatives of the main sorts of creatures that comprise the worldwide fauna of some 2,568 species of damselflies, in 22 different families. The jewelwings represent a relatively primitive group, with extremely dense wing venation and non-stalked wings. The spreadwings are off on a branch of their own, with body colours that are formed either by iridescence or a greyish pigment called pruinosity (at least among the Alberta species). And finally, the pond damsels are the so-called typical damselflies, and they usually possess bright body colours—typically blue or green—as well.

This book deals mainly with adult damselflies, and not their larvae. Larvae are more difficult to identify, although the three families have distinctive larval types. Among the bluets, larvae identification can be so tricky that sometimes only molecular diagnostic techniques can distinguish one species from another. In other words, you have to grind them up and chemically analyze the slurry.

To identify adult damselflies, on the other hand, it is essential to know a certain amount about their structure. Many people think that knowledge of wing veins is needed, but luckily this is not the case for our Alberta species. Body features are much more important. On the head, the main things to look for are the postocular spots—large, pale spots behind the eyes of some members of the pond damsel family that may or may not be joined by a pale bar across the top of the head.

Behind the head, the thoracic "neck" is called the prothorax, and its markings can be important, especially on the upper side, called the pronotum. Behind the pronotum lies the pterothorax (the part of the thorax that bears the wings). Markings on the pterothorax are also often important, and the most frequently consulted of these are the antehumeral stripes, which I prefer to call the pale shoulder stripes. These are the paired pale (usually blue or green) stripes on either side of the dark area down the top middle of the pterothorax. Below the pale shoulder stripe lies the dark shoulder stripe, sometimes called the humeral stripe.

On females the upper front edge of the pterothorax is thickened to form what I call "shoulder pads." In the technical literature the name for these is mesostigmal laminae—an awful bit of jargon if ever there was one. These pads are important in mating, and they are useful for the identification of some species of pond damsels, although they are difficult to see without good lighting and at least 10-power magnification from a hand lens or a microscope.

The abdomen is also an important source of diagnostic features, and in order to locate these accurately you will need to know which segment of the abdomen is which. The segments are numbered from the base to the tip—in other words, starting where the abdomen attaches to the thorax. The first two segments are short, followed by segments 3 to 7, which are long, and then 8, 9, and 10, which are again short. The tops of the abdominal segments are typically marked with black and at least one pale colour in pond damsels, and the exact nature of these markings can be important to identification.

At the very tip of the abdomen of male damselflies, there are two pairs of sexual claspers, one above the other. I like to call them "upper claspers" and "lower claspers," but this is not the usual terminology from a technical standpoint. The uppers are called cerci in modern works, and in older books they may be referred to as superior abdominal appendages. In like fashion, the lowers were called inferior abdominal appendages. The lower claspers are now called paraprocts, a word that establishes their evolutionary connection with the paraprocts of other insects. In my experience, and since their function is to clasp the female, most people prefer the familiar English word "clasper" to any of the former terms. To examine the claspers, it helps to hold the damselfly in your hand and use a 10-power (10X) hand lens.

The pond in Patricia Ravine. Everyone who loves damselflies should find his or her favourite pond, and get to know it well.

2
A Day in the Life of a Damselfly

When I was a boy, some of my favourite books were stories of insect watching. I loved to read Edwin Way Teale's accounts of insect behaviour, written while lounging around on his run-down, vacant-lot property, or Jean-Henri Fabre's brilliant musings on the insects of Provence, in the south of France. It wasn't until high school, however, that I finally decided to set aside an entire day from what seemed like a busy life at the time, and devote it to damselfly watching. At the time, I was a subscriber to *Cordulia*, an amateur journal devoted to odonate and ground beetle observations, published in Quebec. This was my inspiration for choosing damselflies as a focus for the day.

It was a warm sunny morning in late June. I packed a lunch, made sure I had a notebook and a pen, took my binoculars, and set off into the Patricia Ravine, below my parents' house in Edmonton's West End. About a half-kilometre up from where the ravine meets the valley of the North Saskatchewan River, there lies a small, shady, stagnant, and mostly unspectacular beaver pond. That is where I decided to spend the day.

At first, there were few insects to be seen. Before about 10 o'clock, many bugs just simply aren't active in our part of the world. Still, that was when I started finding damselflies in the grassy slope uphill from the pond, warmed by the first rays of morning sunlight over the lip of the ravine. I didn't know at the time that they had already sipped at dewdrops, warmed their bodies in the sunlight, and waited until the air temperature was to their liking.

As I watched, I noticed that they were hovering among the grasses and the low weeds, picking small insects off leaves and stems. Generally, that is how damselflies feed—they hover slowly through vegetation and they strike with their long spiny legs at any off-colour blotches on plant material, in the hopes that the blotches might turn out to be resting edible bugs.

I caught a few of these foraging damsels and marked them with bright orange model paint on their wings. I still have a photo of one, hiding behind a grass stem, as if I couldn't see it—from both sides of the stem, eyes protrude, and behind them the orange-splotched wings as well. From the photos I took that day, I can tell in retrospect that I was watching taiga bluets (*Coenagrion resolutum*). Using paint, by the way, was an unwise thing to do in retrospect, since it may have affected their social standing among others of their kind, or their ability to evade predators. A less obvious mark would have been a better idea.

As the day warmed, the damselflies started moving down toward the pond, and appearing as mating pairs. Near the pond, I watched as males hovered among the grasses and instead of darting at food, darted at other damselflies. If a male encountered a female, he would then attempt to grab her by the neck. This action is unmistakable. The male grabs the female with his legs, around the head and thorax, and then arches his body so that the tip of his abdomen reaches up under his chin and his abdominal claspers take hold of the female's neck (that is, her prothorax). If she is receptive to the male, the process takes but an instant. If she is not, she will struggle and often dislodge the male before he can put the claspers on her. Apparently, some damselfly females are known to reject males by placing their spiny forelegs up over their necks to form what the famous odonatologist Philip Corbet calls a "spiked dog collar."

Once the male has the female in his grip, the two are said to be "in tandem." They often fly together to a low perch at this point. Damselfly mating is an extremely odd process, since the male completes the first step by himself. He transfers sperm from the tip of his abdomen, where his testes are located, to the underside of the base of his abdomen, where he has a second set of sex organs, including a structure that acts as a penis. Once the male has completed his internal business, it is the female's turn to arch her body and swing her abdomen forward. She makes contact between the tip of her own abdomen and the secondary sex organs of the male, and they mate. Inside her body, odd things happen. The male, for one thing, is able to remove sperm from previous matings. This phenomenon was first reported in 1979, three years after my day at Patricia Pond, in the ebony jewelwing (*Calopteryx maculata*), by Jonathan Waage. Since then, other odonatologists have shown that it is a widespread phenomenon among damselflies, involving removal, repositioning, or dilution of the first male's sperm.

This process is made possible because the sperm does not go directly to the fertilization of the egg, as it does in humans. Instead, it is stored in a structure called a spermatheca. Once the female has sperm in her spermatheca, she is able to fertilize her eggs, and fertilization is one of the last things to happen to a damselfly egg before it is laid.

The position that a mating damselfly pair adopts is, in side view, something like a classic Valentine's heart. In fact, there is apparently some evidence that the shape of a Valentine's heart (which looks very little like a human heart) was inspired by red-coloured damselflies in Europe, many centuries ago. If this is true, it certainly makes for an interesting connection between human culture and the study of damselflies, and it puts a nice romantic spin on damselfly mating as well. Although it has been called "the wheel position" for many

Top: A damselfly devouring a baby grasshopper that it snatched from a grass blade moments before. Bottom: The claspers on the end of the abdomen of a particularly well-endowed river bluet.

Top: A little blue heart—the position in which all dragonflies and damselflies mate. Bottom: The male keeps his grip, while the female deposits eggs under the water, presumably fertilized by his, and only his, sperm.

years, I agree with those who feel that the position should be renamed the "heart position."

Most mating damselflies do not adopt this position for long, and this is surely true for taiga bluets. However, the male often keeps his grip on the female's neck for much longer than you might think necessary, and that is indeed what I remember seeing among the taiga bluets at the beaver pond. Together, the pairs flew out over the water and began searching for plant stems either just below the water surface or emerging from it. On these, the female laid her eggs while the male remained attached. The eggs are laid right inside the tissues of water plants, in slits that the female cuts with her ovipositor, a structure that looks like a folded jackknife blade on the underside of the tip of her abdomen. Of course, the male remains with the female until her eggs are laid, for fear that other males might find her in the meantime and remove his sperm.

When a female is resting horizontally on a perch or a bit of floating water plant, the male can hold his entire body straight up in the air by the grip of his claspers alone. This has been called the "sentinel position" by odonatologists, and it apparently allows males to see farther and take off faster than if they perch horizontally. In Europe, Gunnar Rehfeldt has shown that males in the sentinel position can have a significant effect on the survival of their mates (in a closely related species, the azure damselfly, *Coenagrion puella*) by rapidly removing them from the path of attacking frogs. It is also possible that males in the sentinel position are an obvious landmark clue that mating pairs could use to locate good, safe places for egg laying, and at least one odonatologist has argued for this as well.

Now, the Patricia Pond is not a swimming hole (or perhaps it's a swimming hole in the same sense that a sewage lagoon is a scenic wetland), but if it were I might have tried snorkelling in it. I've seen damselflies before while snorkelling in Alberta lakes, since bluets (and apparently the jewelwings as well) often submerge completely while they are laying their eggs. In some instances the male stays with the female while she walks backwards down a submerged plant stem, but in others he stays up top and waits for her to return. She can stay under for long periods of time, since her body becomes coated with a thin film of air. The air serves as a "physical gill" into which oxygen diffuses and out of which carbon dioxide diffuses.

That's about where my observations at the Patricia Pond came to an end, but of course there is more to the story than that. Most female damselflies have the potential to lay thousands of eggs over the course of their lifetimes, but in real life they usually don't live long enough to lay more than about one hundred. Still, that makes them extremely fecund creatures, capable of building up huge populations in very short periods of time (even if only half survive, a single female could give rise to a population of more than six million damsels after only four years).

Usually, when we encounter damselflies we see them either hunting for food away from their breeding areas or mating and laying eggs at the water's edge. These are surely the most important themes in a damselfly's life, but there is one more that is probably just as much so—dispersal. Dispersal is simply the tendency of damselflies to fly to new breeding areas rather than sticking around at the old home pond. It's difficult to know how often or how far damselflies disperse on average, but we can use some indirect evidence to give us an idea. As you read through the species accounts in Chapter 7, pay particular attention to the stories of damselflies that found and colonized small patches of isolated breeding habitat. It is the general impression of many odonatologists that there is a fine and gentle rain of damselflies that falls over much of the world, each and every day in the right seasons. Most of these wayward damsels live out their lives without consequence, but on occasion one or two might give rise to a new population, far from the nearest area occupied by their species. In the technical literature, there are some intriguing observations of damselflies spontaneously flying up into the sky until they are no longer visible and at the mercy of the winds. Forktails seem especially prone to this sort of thing. Some migrate and some don't (within a single population), for reasons that are unclear but might be the result of increasing population density. In British Columbia, my old friend and undergraduate classmate Brad Anholt found that larger boreal bluets were more inclined to migrate than smaller ones.

Damselflies also exhibit some perplexing behaviour patterns, such as wing-clapping. This is easily recognizable and involves the wings suddenly jerking outward and then back together. Apparently, wing-clapping in male spread-wings and pond damsels serves both as a signal to other damselflies (so they recognize the individual doing the clapping as a damselfly) and as a way of cooling their bodies. Both sexes of jewelwings do it, and jewelwing behaviour is described in greater detail in the treatment of our one and only jewelwing species. Damselflies, by the way, are more or less cold-blooded. Maintaining a constant high body temperature is possible only for large dragonflies. Damselflies are therefore much less able to control their body temperatures than are dragonflies, moths, or tiger beetles.

Of course, damselflies also have enemies. The obvious ones are their predators. Adult damselflies are eaten by birds, dragonflies, other damselflies, water spiders, and frogs, just to mention a few types of predators. Larvae are eaten by fishes, water birds, water beetles, dragonfly larvae, backswimmers, and giant water bugs. Damselflies can also have parasites. There are gregarine protozoans (one-celled animals) that affect both adults and larvae and can block the digestive tract and cause death. In Europe, A. Åbro found that 100 percent of adult northern bluets were infected by the time the flight season was in full swing.

The most obvious parasites of damselflies are the bright red water mites (most of which are members of the family Arrenuridae, as I understand it) that attach harmlessly to older larvae as larvae themselves, then reattach to the

A big, mean, scary female water spider—*Dolomedes triton*—one of the creatures that regularly devours unsuspecting damselflies.

emerging adult, near the bases of the damselfly's legs. They feed on the body fluids of the damselfly until it returns to the water after its prereproductive period. Then they return to the water themselves and carry on to become little predators. These parasitic mites can be very hard on the damselflies they use as hosts.

For most of our species, adults appear in late May or early June, and the eggs are laid once the adults are a few days to a few weeks old and mature enough, in the physical sense, to mate and reproduce. The eggs are laid in June and July, and they hatch relatively quickly to become newborn larvae (unless they are consumed by tiny egg parasites in the wasp family Chalcidae). Larval growth is fast, and by the time fall rolls around and the water cools down, most larvae are almost full-sized—at least those that have a one-year life cycle. Those that stay in the pond for two years or more will still be tiny in the fall. When the ice

Top: A typical pond damsel larva, blending in with its surroundings. Bottom left: The angular-headed larva of the western red damsel. Bottom right: A hibernating damselfly larva, frozen solid in the ice, much like the rest of us feel each January and February.

forms, some larvae remain active all winter under the ice, while others cling to the bottom of the ice and allow it to freeze around them. In suspended animation, they spend the winter completely out of reach of predators.

The exceptions to this pattern are the spreadwings, which lay eggs late in the season, above the water line, and the eggs overwinter, usually under snow. Spreadwing larvae develop in meltwater ponds in the spring and early summer, and the adults emerge later in the season than other damselflies, but before their ponds dry up. For this reason, they have no choice but to complete their life cycle in a single season.

When damselfly larvae are ready to emerge as adults, they climb out of the water and onto a favourable perch. Usually, this is a plant stem emerging from the water. The larva's skin splits lengthwise down the back of the thorax, and the adult emerges, spreading and expanding its wings. When the transformation is complete, the still-soft adult takes its "maiden flight" to nearby vegetation and patiently waits while its body hardens and darkens in colour. This is a fascinating process to watch and one that every naturalist should witness at least once in his or her lifetime. Since the adult emerges from a crawling larva that is something, but not much, like the adult, I personally find it even more interesting than watching eggs or butterfly pupae hatch. After all, a pupa has time to rearrange itself internally, while an odonate larva has to go directly from an aquatic, water-breathing stage to a flying, air-breathing stage, all in the space of a mere half-hour, even though most of the internal rearrangement of structures has transpired before this occurs.

The behaviour of damselfly larvae is an interesting subject in its own right. A great deal of scientific study has been devoted to how, exactly, the larvae manage to get food, defend good feeding spots against intruders, and decide where and when not to go looking for other places to live. For the most part, these studies have been performed by people interested in the feeding habits of animals in a broader, more theoretical way and not merely in damselfly larvae for their own sake. For these people, damselfly larvae are a "model system" with which they can test their broader hypotheses.

Personally, I find this approach unsatisfying, since so often such so-called model systems turn out to be unique and quirky, and not a general model that can be applied to all animals. If and when this happens to those who have focused on damselflies, those of us who value knowledge of damselfly larvae for their own sake will benefit greatly, since our favourite organisms will have been the focus of a great deal of time, money, and scholarly attention. And if the damselfly larvae turn out to be a model system after all, well, then they will get more of the glory they deserve.

A lovely pond near a northern peatland—excellent habitat, especially for the Eurasian bluets, and moose.

3
Damselflies and Wetlands

In Alberta, damselflies are intimately linked to wetlands. Hunters think of our province as part of the "duck breeding factory" for North America, because most of the waterfowl on the continent breed in what many people call the "pothole lakes" of the Canadian prairies. Well, these same habitats are home to most of our damselflies. Overall, we have far fewer lakes here than in adjacent Saskatchewan or British Columbia, but still, Alberta is a wet enough place to support a respectable damselfly fauna.

Damselflies in Alberta almost all prefer standing rather than flowing waters. The two obvious exceptions are among our least common species—the river jewelwing (*Calopteryx aequabilis*) and the river bluet (*Enallagma anna*). All the other damselflies in the province are most at home in lakes and ponds. Given the choice, it is also likely that most prefer ponds to lakes as well. The larvae of damselflies generally live among water plants and need emergent vegetation (plants that stick out of the water) in order to moult to the adult stage (usually about 5–10 cm above the water surface). For this reason, deep waters and heavy waves are not to a damselfly's liking. However, calm bays and shallows still support large damselfly populations, even on the largest of our lakes.

Damselflies distinguish their habitats in even fussier ways as well. Some species are found almost exclusively in the absence of fish predators. Others seem more at home with fishes (the official plural of *fish*, if more than one species is involved), and the larvae of these species show behavioural modifications that reduce the likelihood that they will be devoured by such predators as pike, perch, or whitefish. In small ponds and marshes, where large fishes cannot survive, there are often two species of small fishes, the brook stickleback (*Culaea inconstans*) and the fathead minnow (*Pimephales promelas*). It has not yet been shown whether these two species have the same effect on damselfly distributions as do larger pike-, perch-, trout- or (in other parts of North America) bass-family fishes.

As well, it is also clear that warmer ponds support more damselflies than cooler ponds. The beaver pond in Patricia Ravine, described in the last chapter, is surrounded by forested slopes and shaded for much of the day. As a result, it is cooler than an equal-sized pond on flat ground surrounded by fields, and indeed it doesn't support very many species of damselflies.

Some ponds are permanent; others are temporary, or seasonal. Here in Alberta we don't hear much about vernal ponds (*vernal* means "spring" and refers to the fact that they fill with meltwater from winter snows), but the importance of these habitats is becoming more and more apparent in other parts of the country, and the continent. Certainly, these are the sorts of habitats that the spreadwings prefer, and the fact that they lay their eggs in dry vegetation, with faith that the ponds will fill with water in the spring, is evidence that these damselflies are adapted to temporary ponds as well. At the very least, spreadwings take advantage of the fact that most ponds and lakes have predictable seasonal changes in water level—high in the spring and low in late summer or fall.

To make things even more interesting, damselflies also respond in predictable ways to the presence of springs. Typically, certain species are found only in spring-fed ponds, much farther north than they can survive in typical ponds. Most likely, this has to do with the fact that springs keep at least part of the pond open throughout the winter and provide microhabitats in which larvae can survive the cold. In ponds that are not spring-fed, the entire water body may freeze to the bottom, and where it is deep enough to avoid this, dissolved oxygen levels may also plummet during late winter. It is no wonder that not all damselfly larvae can tolerate these conditions.

Those who study springs often classify them as either cold springs or warm springs. This is a fuzzy distinction (based mainly on what the water feels like when you stick your fingers in it), but it does seem to hold some truth with respect to which damselfly species inhabit which spring-fed water bodies. In Alberta, the only species that seems to require hot springs is the vivid dancer (*Argia vivida*), which is known in the province only from the Cave and Basin Hot Springs in Banff National Park. In cool springs, the most typical inhabitant is the western red damsel (*Amphiagrion abbreviatum*), but the Pacific and plains forktails (*Ischnura cervula* and *I. damula*) also show strong associations as well.

We humans have also created an odd sort of artificial spring-like environment as well—the warm water discharge from power plants. Just as manatees spend winters at power plants in Florida, the few power plants that have been examined in Alberta have produced interesting damselfly records as well. In particular, the three coal-fired plants on Wabamun Lake support an extremely interesting population of the plains forktail. The Sheerness Power Plant south of Hannah has produced a wetland that is also worth monitoring for new and interesting records.

Perhaps the most intriguing question regarding damselflies and how they are distributed in their habitats in Alberta is this: are damselflies often trapped

Forktails are filled with wanderlust, and as a consequence they are good at finding and colonizing new habitats.

in ecological islands, like finches on the Galapagos or tiger beetles on isolated sand dunefields, or does the evidence actually point to the opposite situation? Well, ponds and lakes are indeed "islands" of suitable habitat in a "sea" of uplands where damselflies cannot breed. As well, anyone who watches damselflies will notice that they stay low to the ground and close to home most of the time. It would seem reasonable to assume that they are not terribly good at dispersing from place to place.

However, we know two things well about the history of wetlands and damselflies, especially in southern Alberta. First, the number of wetland areas on the prairies has increased dramatically in historical times, despite the filling and draining of shallow sloughs. Dugouts for cattle watering, irrigation reservoirs, waterfowl habitat projects, and the like have all changed the face of the province, at least from a damselfly's point of view. Second, many species of damselflies that are now well known from these newly created wetlands were not recorded during the early years of damselfly study in the province, despite extensive collecting of specimens by eminently qualified people. From this, it is reasonable to conclude that damselflies are good enough at dispersing that they can easily find and occupy new habitats as they become available. In part, this is why the study of damselflies from an ecological point of view is such an interesting one—nothing stays the same for long, and the damselfly "deck" is continually being reshuffled.

A fine portrait of Francis Whitehouse at the top of his game—sportsman, banker, poet, novelist, and the father of damselfly studies in Alberta.

4
The History of Damselfly Study in Alberta

My own interest in insects began with butterflies and beetles, and that's how I became connected with the study of damselflies. I fell in love with tiger beetles at the age of 13, and on the inside cover of my favourite book, J.B. Wallis's *Cicindelidae of Canada*, there appeared an advertisement for *The Odonata of Canada and Alaska*, by Edmund Walker and Philip Corbet. As a junior high school student, I couldn't afford this three-volume, $100 set of books, but I eventually acquired a free publication from the Government of Québec, *Les Libellules du Québec*, by Adrien Robert. Armed with my typically Albertan quasi-understanding of French, I set out to identify dragonflies and damselflies.

It was easier than I thought it would be, since many terms in entomology books are common to both languages. Soon, I added Rob Cannings' and Kathleen Stuart's fine book, *The Dragonflies of British Columbia*, to my bookshelf (an almost-impossible-to-find treasure nowadays) and then, eventually, the volumes by Walker and Corbet.

Two things seemed clear to me back then. First, British Columbia was very well known in terms of its odonates. Second, Alberta was not. After a couple of seasons spent collecting specimens and encouraging my friends (and especially Felix Sperling, who is now the systematic entomology professor at the University of Alberta) to do the same, I published a small paper on my discoveries and roughly doubled the number of published odonate localities for the province. I was quite proud of this, as well as of the title of the journal in which my first "scientific paper" appeared: *Notulae Odonatologicae* (although all that means is "notes on dragonflies and damselflies" in Latin). That paper, in retrospect, was mainly of value because of one interesting speculation (see under the western red damsel) and one clumsy mistake, which I had to discover eighteen years later for myself (see under the eastern forktail).

The impression I had back then, and no doubt passed on to others, was that our predecessors were dabblers, who really didn't take the time to fully document the odonate fauna of Alberta. After all, just about everyone who became interested in these insects during the 1990s found at least one new species for the provincial list. I was the first to report river jewelwings, alkali bluets, tule bluets, and western forktails; Natasha Page found the first plains forktails; and Christine Rice and Jon Hornung found the first eastern forktails. Those who did not get first records for the province got seconds or thirds. Surely, we figured, we were pioneers in an unexplored land.

Slowly, however, this impression has crumbled in the face of evidence from the past. I first began to sense this when a fellow named Gavin More started collecting specimen records of Alberta odonates as part of a contract with Parks Canada to determine their conservation status. He uncovered large numbers of specimens in many major museums (not many in Alberta, mind you)—many more than I expected on the basis of published localities in Walker and Corbet.

Then I began to learn more about Francis C. Whitehouse. In 1918, Whitehouse published a booklet entitled *Dragonflies (Odonata) of Alberta* through the Alberta Natural History Society in Red Deer. I had known of this booklet for some time, but it didn't impress me much at first, what with its crude, crowded drawings and badly aligned tables. Whitehouse, however, was in fact a very fine amateur odonatist. He was also a poet, a novelist (he wrote *Plain Folks: A Story of the Canadian Prairies*), and a writer of books on sport fishing. His interest in fishing was apparently what led to a passion for damselflies and dragonflies, and his dragonfly booklet begins with the following couplet, presumably one he wrote himself:

> *A bug designed a double debt to pay,*
> *To feed the fish and keep the flies away.*

Whitehouse was born in 1879 in Leamington Spa, Warwickshire, England. He arrived in Canada as a 26-year-old in 1905, the year Alberta became a province. While here, he managed the Bank of Commerce in Red Deer. Whitehouse also spent time in British Columbia, Saskatchewan, and Manitoba, and was no doubt broadly familiar with the damselflies and dragonflies of western Canada. His interest in odonates was fuelled by his connection with Edmund Walker in Ontario, and he published papers on the dragonflies of Alberta, British Columbia, and Jamaica. He was also interested in butterflies and moths, and was a member of the Alberta Natural History Society, which had its headquarters in Red Deer at the time.

Apparently, he was also involved in the Boys and Girls Clubs in Red Deer and he founded the first "Pig Club" (a forerunner of 4-H) in 1918, the year he published his dragonfly booklet. I can't help but wonder how much more we would know about Alberta odonates if Whitehouse had succeeded in inspiring a few young entomologists rather than aspiring hog farmers! In retrospect, if I had known of his booklet when I was a child, I may well have become an

Top: Francis Whitehouse (back row, right) with what may well be the Red Deer Pig Club. Bottom left: Rob Cannings, the legendary British Columbian odonatologist. Bottom right: Gordon Pritchard, the finest academic odonatist in Alberta's history.

inspiree of Whitehouse myself. I didn't even come close to meeting Whitehouse, since he retired to Phoenix, Arizona, where he died in 1959, the year after I was born. Whitehouse's odonate collections now reside in museums in British Columbia and Ontario, although some specimens remain in the E. H. Strickland Entomology Museum at the University of Alberta, while others have been spread around the world.

Edmund Walker, Whitehouse's Toronto colleague, was one of the most important entomologists in Canadian history. Amazingly, his university training in biology consisted of only a bachelor's degree, although he did complete his training in medicine as well. The tradition of largely or completely self-taught biologists was strong in those days (remember, it also included Charles Darwin), and in my opinion it reminds us all that the PhD is not necessarily the measure of all things entomological.

Walker not only wrote the three-volume masterpiece *The Odonata of Canada and Alaska*, but also discovered and named an entire order of insects, the rock-crawlers or Notoptera, which he found, in 1913, in Banff National Park during one of his western expeditions. Walker was a professor of zoology at the University of Toronto at the time, and in 1918 he joined the curatorial staff of the Royal Ontario Museum. In his brief autobiography, Walker mentions coming west in 1921, in a rickety Model-T Ford with two other colleagues, and visiting "my old friend Frank Whitehouse" in Nelson, British Columbia. By that time, Whitehouse had left Alberta and was working on his very impressive paper on the odonates of B.C. In 1925, Walker honoured Whitehouse by naming the dragonfly *Somatochlora whitehousei* in his honour. We now call this species the Whitehouse's emerald in English.

Walker's last volume on the Canadian dragonflies was completed after his death by another stellar odonatist, Philip Corbet. Corbet visited Alberta a number of times and tells me that he has fond memories of the entomology department at the University of Alberta and its tradition of hospitality toward visiting academics. In 1963 he accepted, but then declined, a one-year term as a sabbatical replacement for the University of Alberta entomologist George Evans. Now, he is best known as the author of *Dragonflies: Behaviour and Ecology of Odonata*, the most comprehensive book ever written on the subject. He currently lives in Cornwall, England, and is still actively involved in the science of dragonflies and damselflies. I had the pleasure of meeting him at a recent workshop on the status of Canadian odonates, and I can attest to his incredible knowledge, meticulous self-honesty, and gentle but commanding nature.

A slightly younger man was to become the principal academic odonatologist in Alberta history. His name is Gordon Pritchard, and I find it interesting to contrast his more typical academic career with the faunistic studies that came before him. As a university researcher, Pritchard has been dependent on grant funding in order to pursue his interest in dragonflies, and funding agencies have long ago decided that "mere faunistics" is not innovative enough to justify their support. As a consequence, Pritchard has focused much more tightly on specific aspects of damselfly biology rather than the sort of broad picture that characterized his predecessors, and people like me. As Pritchard explained it to me, "Although I get really excited about some things, I'm not like you, where I get excited about everything." Gordon's nature is to look at details and mechanisms, and that is also the nature of university-based biology.

This should not, however, give the impression that university professors are any different from the rest of us, under the skin. Gordon Pritchard grew up in the English Midlands, and he still remembers the first odonate he ever caught—a big brown hawker (*Aeshna grandis*). He got it with his jacket at age nine or ten, when he was attending Burton-On-Trent Grammar School. Gordon was inspired by his biology teacher and wanted to be a biology teacher himself when he grew up, but instead he entered the entomology program at Imperial College in London, following a number of others who had been inspired by the same instructor.

At Imperial College, Pritchard spent about a third of his time playing rugby, another third playing drums with a jazz band, and the final third in class. To hear him tell the tale, this was not a terribly inspiring time entomologically. The system had taken his childhood love of insects and transformed it into a desire to make a living spraying pesticides at pests. At the time, however, one way to avoid conscription into the army was to become a graduate student, so Pritchard continued his studies rather than jump straight into the "nozzle-head" business.

Since he was a good student, Pritchard was in the running for two scholarships, one to stay at Imperial College and one to go to the University of British Columbia. Choosing between the two was not easy. As Pritchard tells the story:

> *I went for an interview before this illustrious panel, of which "Wiggy" [Sir Vincent B. Wigglesworth] was the chairman. And my only recollection of this was that I was down in this pit and these people were sitting up there at the high table looking down at me and asking all these erudite questions. I stumbled through that, and then just before the interview was closed Wiggy looked down over his glasses and said, "I see, Pritchard, that you've also applied for a Commonwealth Scholarship to go to Canada. Now supposing that you were offered both of these, which one would you choose?" So quick as a flash I said, "Well, of course I'd stay, sir—I'd take the ARC Scholarship and stay here." And there was this pause, and then he looked at me and he said, "You know, Vancouver is an awfully nice place." So I slunk out of there, and in actual fact I was offered both scholarships but wound up not at the University of British Columbia, but at the University of Alberta.*

At the University of Alberta in Edmonton, in 1960, Gordon became one of eight graduate students in an entomology department with only three faculty members: Brian Hocking (the chairman), George Evans (an ecologist, and Pritchard's supervisor), and George Ball (a systematist). Two of his contemporary students were the tiger beetle specialist Rick Freitag and the now-famous biodiversity expert Terry Erwin of the Smithsonian Institution. They had coffee twice a day with zoologists and Canadian Wildlife Service employees, and there Gordon met such well-known biologists as Ralph Nursall, David Boag, John Holmes, and Don Ross. "It was always wonderful—with ideas being bashed around—you really felt part of an exciting group."

Gordon did his PhD research near Flatbush, Alberta, where Brian Hocking had made a deal with a man called Colonel Hughes. Hughes owned a farm that also served as a field station (before the department had its own permanent field station at George Lake). Gordon stayed in a shack with another student, David Haphold, who worked on mosquitoes. It was an isolated place, there was nothing else to do, and it was home to a good fauna of dragonflies. Gordon spent only one summer there, working in a lab made out of an old army truck—a great lumbering vehicle with a tarp on top to prevent leaks. Inside, Gordon filmed the predatory strikes of larvae, and the research was completed quickly and efficiently. As Gordon recalls, he and David virtually subsisted on rabbit stew, made from snowshoe hares shot by the Hughes boys.

Gordon finished his PhD in 1963 and tried to live up to the terms of his scholarship by going back to England to become an economic entomologist. But the people in Britain were not very welcoming at a distance, so Gordon took a job doing Queensland fruit-fly studies in Australia, which he abandoned after only a few years. That was when Brian Hocking offered to let him replace George Evans while Evans was on sabbatical leave—the one-year position that Philip Corbet had accepted and then declined. So Gordon came to the University of Alberta for a year, and in the meantime got an offer from Calgary in 1966, starting work there in 1967. The head of the biology department in Calgary was Jim Cragg, who had been an external examiner at Imperial College. Hocking had also spent time at "I. C.," demonstrating further how much entomology in Alberta owed its roots to England at the time.

Once on staff in Calgary, Gordon began work on the population ecology of crane fly larvae at the Kananaskis Experimental Station. In fact, he spent ten years working almost exclusively on crane flies. Then, in the late 1970s, he turned back to odonates, starting with vivid dancers at Banff. He then embarked on a comparative study of the life history of this species in various springs, which he found partly with the help of *Hot Springs of Western Canada*, a quaint little bathing booklet published by the Labrador Tea Company. This took him all over the western United States at a rate of about a state per year.

Finally, in the late 1980s, when his work was taking him pretty far south, he realized it was time to go to the tropics, "where dragonflies really come from."

One of Gordon's most frequently cited conclusions with respect to odonates is that they retain their tropical temperature responses and are thus "prisoners of their tropical past." Gordon then went to Costa Rica to study other members of the dancer genus *Argia* and found that they were a "taxonomic nightmare." So he switched to the rubyspot damselflies (*Hetaerina*) and did some work on the tropical genus *Cora* as well. These he describes as "wonderful creatures," especially notable for the gills that the larvae possess on the sides of their abdomens. This led to work on the American rubyspot (*H. americana*) near the northern edge of its range, where it lives in hot springs in North Dakota (a species that might someday show up in southeastern Alberta, although I don't know of any hot springs that could support it).

Of course, Gordon was not just a graduate student and a researcher, he was also a professor and a supervisor. In his lab, the odonatological students included Kelvin Conrad, Jack Zloty, and Mark Leggot. Leggot worked on vivid dancers and their thermal characteristics, and is a bibliophile who loves libraries. He is now a librarian at the University of Manitoba.

Jack Zloty, a taxonomist whose original love was hover flies, studied physical education for his bachelor's degree in Poland. He arrived here in Alberta as a refugee, interested in insects in general. Gordon Pritchard took Jack under his wing, at which time Jack organized the University of Calgary's hover fly collection and then started a master's degree program, working on *Ameletus* mayflies. Then Jack went to Costa Rica and helped Gordon associate rubyspot larvae and adults using a biochemical identification technique called electrophoresis. For his PhD, Jack was going to look at the evolution of the dancer genus in North America using electrophoresis and DNA analysis. Gordon and Jack had a couple of great summers collecting material for this project, and the specimens were frozen in a huge walk-in freezer on campus in Calgary. Then one weekend in February, as Gordon put it, "the bloody freezer not only stopped working, but turned into an oven, and went overnight from minus 80 to plus 80." The next summer they re-collected a great deal of the material needed, but by then the "gloss had gone off it," so Jack switched back to *Ameletus* mayflies for his PhD. Jack is now a consultant and keeps up his interest in entomology, often in collaboration with Pritchard.

Kelvin Conrad, Pritchard's third damselfly student, also became involved in the study of vivid dancers. He considered working in Banff, "but the paperwork was overwhelming and the chance of 'demonic intrusions' by tourists was too great for any long-term work." So he focused instead on two springs in British Columbia. The plan was to compare the damselflies of Albert Canyon (with a two- to three-year life cycle) and Halcyon Hotsprings (with a one-year life cycle), and relate differences in demography and behaviour to differences in life cycle.

However, like Jack Zloty, Kelvin was not exactly blessed with good fortune during his time at the University of Calgary. First, he missed the first month of

Top: A modern-looking, normal-sized fossil damselfly wing from new Red Deer, about 55 million years old.
Bottom: Jon Hornung (left) and Christine Rice (right)—two graduate students at the University of Alberta who made damselflies part of their thesis studies.

his subjects' flight season, and then had to return to Calgary before the season was over. On the basis of other studies, he had expected the damselflies to be on the wing exclusively in July. Then, as he tells it:

> *I was camping at Canyon Hotsprings...for about a month. It had been raining for four days and I was glad for a chance to dry out. The study site was a mineral seepage swamp in the middle of a clear-cut, and CN/CP were dynamiting through cliffs to twin the railway nearby. A furry, red-headed fellow dressed in green pulled up in a green 4 x 4 and told me the government had closed the forest and I had to leave because I was a forest fire hazard. I protested loudly, standing there in the middle of a swamp where there wasn't any forest, but that ended my field season.*

Kelvin finally settled on a study at Halcyon Springs of how the mating system of vivid dancers maintains more than one colour morph among the females. Kelvin is still very proud of this study, and after finishing his MSc he went on to do a PhD at Queen's University, studying birds. He now lives in England and has married a fellow odonatologist, Joanna Freeland. In fact, I just sent her some samples of green darners (*Anax junius*) for a DNA study.

In nearby Saskatchewan, another entomologist has also contributed significantly to the understanding of our Alberta damselflies. His name is Cedric Gillott, and he too was originally from England. The way he tells his story:

> *Soon after I came to Saskatoon (in 1965), I decided that I would like to change my research direction and experimental insect (I'd worked on locusts for my PhD), so went back to a long-standing interest in aquatic biology...The damselflies piqued my interest and I became keenly interested to learn how they overwintered in sloughs and lakes covered with several feet of ice. Fortunately, a good student came along—Bill Sawchyn—and NSERC generously gave me an equipment grant to purchase 4 growth chambers with temperature and light control. Bill had a thorough knowledge of the prairie slough ecosystem, having done a MSc with Ted Hammer on Diaptomus [a crustacean]. We were also fortunate to have the great help of the late Norman Church who had expertise in diapause, having worked on wheatstem sawfly. Bill prepared a very good thesis, which was examined by Phil Corbet, then at Waterloo University, I believe...Unfortunately, the work was not carried on, in part, I think, because of its seasonal nature and also because as other students came along, I found myself focusing more on the insect endocrine system, and its role in growth and reproduction. Gradually, in fact, I went back to working on Orthoptera, namely the migratory grasshopper (Melanoplus*

sanguinipes), *specifically and curiously the "non-diapause"(!) strain developed at the Agriculture Canada Research Station.*

Work on Alberta damselflies and dragonflies was not limited to the modern fauna either. In the Red Deer area, there are significant fossil deposits from the Paleocene period (dating from just after the extinction of the dinosaurs), and in these the fossils of odonates are sometimes found alongside those of ancient plants and fishes. The late Dennis Wighton, then the administrative officer for the Department of Genetics at the University of Alberta, took a keen interest in these fossils and published on them, often with the aid of Mark Wilson, a University of Alberta professor with interests in fossil fishes and insects.

Various other people engaged in short-term studies of particular odonate faunas, and these have been important contributions as well. In 1980, Graham and Dierdre Griffiths wrote a consulting report on the odonates of the Canadian Nature Federation's Clifford E. Lee Nature Sanctuary, a very typical central Alberta shallow marsh. Godo Stoyke, an environmental educator, published on the odonates of the University of Alberta's Devonian Botanical Gardens and also worked on the fauna at Ministik Lake. Unfortunately, all of Godo's field notes were lost for the latter study, although his memory remains fresh on much of the detail.

During the mid-1980s, Donald Hilton studied the Cypress Hills odonate fauna. Don was born in Calgary and is a PhD graduate of the University of Alberta, where he studied ground squirrel parasitism for his doctorate. Intriguing as that subject must be, Don was inspired by Dr. George Ball to take on a side interest in beetles that eventually transformed itself into an interest in peatland odonates, which Don first fell in love with while studying the insects in pitcher plants. He is now a professor and chair of the Department of Biological Sciences at Bishop's University in Quebec. However, he comes back to Alberta often and especially enjoys the Cypress Hills, where he has also studied bumblebees, earthworms, and butterflies. Don is an active member of the Alberta naturalist community and often contributes a column entitled "It's a Small World" to the Federation of Alberta Naturalists' journal, *The Alberta Naturalist*.

Carroll Perkins, a superb insect photographer, has been prowling the wetlands around Edmonton (and especially Elk Island National Park) for decades now, and as a result his odonate photo collection is one of the best in the province. A number of his slides were used in this book, and I'm sure you'll agree it's easy to spot them at a glance—they are that good! As well, at the Wagner Natural Area, the summer student in 1995 was Natasha Klingsh (now Natasha Page), and Natasha decided to survey the odonates of this interesting peatland as part of her summer duties.

Another undergraduate student, Christine Rice, undertook a similar survey at the Beaverhill Lake Bird Observatory in 1998. Christine is now working on

The first meeting of HOGSOC, the "Hyperboreal Odonatists' Guild and Social Club," with Carroll Perkins, Kamal Ghandi, John Acorn, Natasha Page, Carole Patterson, and Dena Stockburger.

her master's degree under the supervision of Lee Foote (of the Department of Renewable Resources at the University of Alberta), studying the effects of cattle-grazing on odonates in prairie wetlands in Alberta. Her boyfriend, Jon Hornung, is also a student of Lee Foote's and a keen odonatist.

Finally, I should mention the contributions of Ed Fuller, who has been collecting Alberta odonates for many years and sending them to Carl Cook, a keen odonatist in Kentucky, in return for click beetles. Ed is a click beetle specialist and completed his PhD at the University of Alberta. Carl has been very good about sending us lists of the odonates he identifies among Ed's specimens.

Those of us who are currently active in the study of Alberta odonates have formed a small, highly informal study group, which we call HOGSOC—the Hyperboreal Odonatists' Guild and Social Club. The group keeps us in touch, reminds us to spend time together in the field as well as in the off-season, and has also inspired some of us to read Philip Corbet's *magnum opus*, one chapter at a time, as a discussion group. If you find that your own interest in damselflies reaches the level where it might be considered "serious" (perhaps *sincere* is a better word), please feel free to contact me and I will include you in HOGSOC's sphere of influence.

What could be more enjoyable than watching damselflies on a sunny day in Alberta?

5

How to Study Damselflies

The study of damselflies, or at least their adults, is a seasonal thing in Alberta. Here, damselflies start flying in May or sometimes June if the spring is a cool one. The pond damsels are most abundant in June and July, followed by the spreadwings, which are most abundant in July and August. After that, things slow down, and the last damselflies disappear somewhere late in September or early in October, depending on the weather and on where you happen to be.

Damselfly larvae, on the other hand, can be studied at any time of year (although spreadwing larvae are not present in winter) so long as you are able to find them in open water, under the ice, or within the ice itself. Unlike the adults, larvae can be kept quite easily in aquariums, and fed on small live food items, the way one feeds the fussy species of tropical fishes (with brine shrimp, black worms, *Daphnia*, and so on—ask about these foods at specialty aquarium shops). If you give your damselfly larvae sticks or plant stems to crawl out on (as well as underwater sticks or vegetation so they can establish individual hunting territories), they may emerge as adults for you. It is important here, however, that you make sure that their maiden flight does not end in a collision with the lid of the aquarium, and a drowning death back in the water. Instead, let them fly to a window, where you can search for them at least once a day.

I suspect that most people reading this book will want to limit their studies to adult damselflies in the field. For this activity, you can take two quite different approaches, depending on whether you are you going to capture them or not. If you are, then you need a net (see Appendix 3 for sources of equipment). Damselflies can be handled gently by the wings. In the hand, it is easy to examine the fine details that are often required for confirming your identifications, using a 10-power hand lens. Properly handled, hundreds of damselflies can be captured, examined, and released in a single day, without harming any of them. Keep in mind that freshly emerged damselflies are soft, shiny, and greenish, in which case you shouldn't catch them at all since they are easily damaged. Their wings in particular have a tendency to stick together and become twisted. Once you have seen one of these "teneral" damselflies, it will be easy for you to recognize them at a distance.

Hand lenses come in a variety of sorts. I prefer a 10-power (10X lens of the "triplet" sort), which I carry on a strap around my neck. When I catch a damselfly, I first hold the lens to my eye and then bring the bug in close enough to be in focus, making sure that the sun illuminates whatever I am looking at. Some people use 20X lenses, or 5X lenses for this purpose, but I find 10X to be about right. In a pinch, you can also use one-half of your binoculars held backwards for in-the-hand identifications as well. The big round magnifying glasses that most people think of when they hear the term are generally not powerful enough for damselfly work, although you might want to carry one if you are working with kids, since they are so much easier to use.

Catching a damselfly with a net is usually a simple matter of sweeping it into the net with a sideways swing, followed immediately by letting the net bag fold over the rim so the insect does not escape. Some of the more nervous damselflies, especially in open areas, may have to be stalked and pounced upon, but most are not aware of your presence until it is already too late. If you find yourself without a net, you can always try catching them with your hands, but believe me, this requires kung-fu reflexes and a gentle touch. Kids often bring me damselflies in proudly clenched fist, and these specimens are usually dead or crippled when the child's fingers uncurl. If you have a hat, or a towel, these are much better than fingers when it comes to emergency damselfly catching.

I have assumed in the above discussion that you will be practising catch-and-release. Collecting damselfly specimens has never had the same appeal as butterfly or even beetle collecting, mostly because the colours of damselflies fade to a dull brown once they die. Still, it can be useful to have a few reference specimens on hand, and I'm sure that a minority of readers of this book will want to pursue this activity for one reason or another. If you choose to collect damselflies, here's how it is done. Place the damselfly in a small, glassine envelope—the kind that are sold for stamp collecting. Then place the envelope in the freezer, which is generally considered a humane way to kill a damselfly. Once the insect is dead (15 minutes will do it), and without delay, immerse it, still in the envelope, in acetone. You can buy acetone at hardware stores, and please do pay attention to the warnings that tell you how flammable it is and how harmful its vapours can be.

After 24 hours or so, take the damselfly out, let it dry for about half an hour, and then place it in a standard odonate storage envelope (again, see Appendix 3 for sources of supplies). In the envelope, place a file card on which you have written the date and place the specimen was collected, your name, and whatever other details you might deem important. Store the specimens in an airtight container and protect them from larder beetles, which will gladly devour them given half a chance. For this purpose, you can use naphthalene (mothballs), paradichlorobenzene (the white stuff that they put in public washroom urinals), or cut up pieces of Vapona pesticide strips.

Top: To properly identify many damselflies, it is necessary (and fun) to examine them in the hand, up close. Bottom: Of course, the damselflies also like to watch us.

Top: Céline Sirois uses a close-focusing monocular—a great piece of equipment for this kind of nature study—to get a look at a male bluet. Bottom: Properly labelled specimens, in a transparent envelope, stored in a pest-proof container, with their colours preserved by soaking in acetone.

CANADA: Alberta
Canal north of Brooks
50° 36.940' N
111° 53.828' W
741 m
July 6, 2001
John A. Acorn

Enallagma ebrium ♂

This is how professional research collections are assembled (although some are pinned, in the usual entomological fashion), and you should always think of your personal collection as part and parcel of the overall, shared resource that insect collections represent. If other serious damselfly enthusiasts ask to borrow specimens or examine your collection, please allow them to do so. Most of us agree that the only valid reason to collect insects of any sort is to use them in the context of research and education, and luckily few people are eccentrically possessive about damselfly specimens, the way they sometimes are with things like butterflies.

If you are concerned about the ethics of damselfly collecting, I sympathize. However, I also suggest strongly that you remain open to the evidence and skeptical of the notion that all collectors are ecologically irresponsible by definition. As you will surely find out as you spend time with damselflies, damselfly populations are usually very large, and damselflies themselves have immense reproductive potential. Butterflies have similar life history characteristics, and even the rich literature on butterfly science contains no clear scientific evidence that collectors represent a threat to population levels. If you are concerned for the well-being of damselflies, you can do more good by focusing on the real problems—habitat destruction being the first and foremost, at least in some parts of the world. See the next chapter for a more extensive discussion of damselfly conservation in Alberta.

For those interested, the Dragonfly Society of the Americas has produced a set of guidelines for odonate collecting, which can be viewed on the web at http://www.afn.org/~iori/oincolgl.html. I belong to this organization, and I recommend it to all of you with an active interest in damselflies and dragonflies.

If you choose not to use a net, you must learn to identify damselflies at a distance. Luckily, damselflies are long enough to qualify as big bugs and their colour patterns can be seen some distance away, especially when one uses binoculars. Different makes and models of binoculars, however, come into focus at different distances. Some will allow you to examine a damselfly while it sits on your feet, while others will not focus any closer than 5 or more metres away. I prefer close-focusing models for damselfly watching, and my favourites so far are the Bausch & Lomb Elites (I use the 10 x 42 model, but the other Elites are good too), or the Eagle Optics Ranger Platinum Class 6 x 32. Mind you, at the closest focal distance, these binoculars (and all others I have examined, by the way) will only show the damselfly through one eye at a time, although the inner workings of your brain may hide this fact from your conscious mind. Although I generally throw away the front lens covers for binoculars, I do keep the left front cover for my Eagle Optics Ranger, and I attach it to the binocular when using the closest focus, effectively transforming it into a monocular, albeit temporarily.

Speaking of which, for those who want a really, really close look at damselflies in the field, I have two suggestions. First, try an ultra-close-focusing monocular. There are a number of models to choose from, although they seem to come and go quickly in manufacturer's catalogs. Right now, Zeiss, Minolta, and Nikon are the only manufacturers I know of with models on the market, but Swift (model 777) and Bushnell (model 14–8200) have produced fine monoculars in the recent past and these models might still be available here and there. Most of these specialized monoculars focus to about half a metre or less, which is very, very good for damselflies. Of course, they only work with one eye at a time, and I must admit that I find them difficult to aim in the field. For some reason, they do not become an intuitively satisfying extension of your normal vision and the switch from naked-eye viewing to monocular viewing is much more jarring than the switch from naked eyes to binoculars.

For this reason, I prefer the second option for ultra-close-up damselfly viewing, the "Carroll Perkins Gizmo." Carroll, our local optics guru here in Edmonton, showed this to me years ago, and it is one of my favourite field techniques. All it involves is a pair of reverse-porro-prism binoculars (the ones with the front lenses closer together than the eyepiece lenses) with a Nikon 5T close-up lens held in front. The reverse-porro design is important, since it allows both visual fields to overlap and can accommodate a single 5T lens in front of both front binocular lenses. The 5T is also important, since it is a high-quality "two-element" lens, and not a blurry single-element lens. With this arrangement, you have to sneak up on the damselfly and lean into focus with your entire body, but the rewards of seeing the damsel with both eyes at once, filling your visual field, are immense. I have even used this apparatus to examine the claspers of male damsels in the field, something that is darned-near impossible any other way.

Using an accessory close-up lens with your binoculars makes it more difficult to switch from distant to nearby viewing. I suspect this is the reason most people opt for the more convenient but less satisfying option of purchasing a close-focusing binocular. I have demonstrated the Carroll Perkins Gizmo on television, and at many nature festivals and public appearances, and I have been dismayed by how few of my fellow naturalists use this marvelous invention.

Regardless of the optics you chose, remember to move slowly and smoothly while watching damselflies, or you will spook them. Sitting in one place is often a good idea. I like to plunk myself on docks, boats, or lawn chairs, or explore wetlands in a float-tube (the floating inner-tube seats that fishermen use) or in chest waders. Once you know what to look for, you can immerse yourself for hours in the lives of these delicate creatures. You will also find that moving slowly or sitting in one place enhances your awareness of the other creatures that damselflies often interact with—such things as stickleback fishes, dragonflies, frogs, spiders, and the various tiny creatures that make up the damselfly diet.

Once you master these simple field techniques, you will be in a position to make a contribution to the study of damselflies in Alberta. In this context, there are many things you can do. For one, we are still guessing at the exact geographic ranges of almost all of our species. The overall patterns seem well-enough established, but the fine details are still missing. Thus, any records of damselflies identified to the species level at particular locations will be of value. Many individuals and organizations are building insect record databases these days, but the one I am most closely connected to is the E. H. Strickland Entomology Museum at the University of Alberta. I can promise that records will be happily received there and freely shared with others. The most important thing, however, is not where you send your records but that you keep records in the first place.

It is also worth noting that damselfly ranges can and do change over time, and tracking the changing distributions of our local species is also of interest. As you read through the species treatments in Chapter 7, you will see over and over where particular damselflies seemed to have spread across the province in recent times. With enough people out watching them, these sorts of movements can be much more convincingly documented. It is my hope that we will soon have enough sophisticated observers in Alberta to provide the sort of finger-on-the-pulse monitoring that we now have among butterfly people. In other places (Britain and New England come quickly to mind), there are indeed enough damselfly enthusiasts to keep good tabs on the local fauna from year to year.

For those with a yearning to do something more involved than merely finding and identifying damselflies, there are an uncountable number of different ecological studies that one could take on as well. Hopefully, some of the accounts in this book will suggest where the holes are in our knowledge. One sure-fire way to generate a useful study would be to choose a local wetland and survey it for damselflies every week or two throughout the field season, keeping track of how many of each species you encounter. That way, you will not only collect a great number of simple occurrence records, but also a record of the adult life history pattern that can be compared to others from other locations. And of course, the more unusual your location, the more interesting your study will be. A shallow wetland near Edmonton or in the southeastern prairies would not be likely to produce new information, since many such areas have been surveyed already. However, a boggy pond in the mountains, a lake in the shield country of Alberta's northeast, or a spring-fed pool in the boreal forest would be fascinating. As well, much can be learned from continual monitoring of new habitats, such as the "lakes" that are built to enhance new subdivisions.

In the last chapter, I paid tribute to just about all the serious odonatists who have ever worked in Alberta. Given the vastness of the province and the immensity of the subject of damselfly biology, the more of us there are out there studying them, the more chance we have of answering the questions they pose for us. I don't want to give the impression that I expect every reader of this book to become an active part of this process, but in an ideal world I also wouldn't complain if that came to pass.

This sedge sprite has tackled and killed a mosquito, but the protection of people is probably the last thing on its tiny, opportunistic mind.

6
Damselfly Conservation in Alberta

If you want to skip this chapter, here's the short version: Alberta's damselflies seem to me to be in pretty good shape, at least as I write these words. But if you want more detail, please read on.

The first thing that anyone should know about damselfly conservation is that it is based on damselfly appreciation. To quote Philip Corbet from the pages of *Dragonflies: Behaviour and Ecology of Odonata*:

The most pervasive influences of dragonflies on people…are esthetic and scientific. Dragonflies provide inspiration for art and poetry, and are widely admired for their beauty and elegance by people who are not necessarily biologists, and of course…they constitute valuable models for the comparative zoologist, especially the student of behavior or ecology.

In Alberta, this is surely true, although it is possible that we haven't quite yet seen our fair share of homegrown odonate art and poetry. Of course, one could argue that damselfly larvae are important aquatic predators, as well as serving as food for various fishes, and that the adults play a similar role on land. This is true, but I am personally leery of fully endorsing the popular belief that "they eat huge numbers of mosquitoes" and that this is their raison d'être. They do eat mosquitoes, to be sure, but it is clear that they also eat a vast variety of other insects and that other factors, such as the weather, seem to have a much greater impact on mosquito numbers. Without entirely abandoning the notion that damselflies and mosquitoes are mortal enemies, I have to confess that I've never seen any scientific evidence supporting the notion that odonates of any sort are a major factor in natural mosquito control here in Canada. I am therefore skeptical that there are things we might accidentally do to kill off the odes and let the mosquitoes run wild. On the other hand, some tropical odonates do seem to play an important role in mosquito control, as one might predict.

In cases like this, a bit of common sense biology can help. Mosquito larvae, for the most part, live right at the water's surface, unlike odonate larvae which prowl on underwater plants and the bottoms of ponds and lakes. Thus, damselfly larvae may not encounter mosquito larvae as often as they encounter other sorts of food. As well, other sorts of mosquito control practices may inadvertently target damselflies. In fact, one Hawaiian damselfly species became seriously endangered when a primarily surface-feeding fish (*Gambusia*, the mosquitofish) was introduced for mosquito control and turned out to be a pretty good predator on damselfly larvae as well. Here in Alberta, as if to make sure nothing is straightforward or easily predictable, this same fish lives alongside the vivid dancer in Banff, without apparent ill effect.

Damselflies have also become popular as "indicators of wetland health." Most environmentalists like the sound of this phrase, since it shows a commendable sort of sensitivity to ecological complexity, and to the little creatures (i.e., things other than ducks and sport fishes). However, those biologists who have considered this issue in detail have trouble, in my opinion, expressing just exactly what damselflies can "indicate" other than the presence of their own species in a particular pond or lake. Each species of wetland plant and animal has its own unique habitat preferences, and each finds and chooses wetlands in its own way. A wetland that is good for damselflies will not necessarily be good for all other species of living things, and a wetland without damselflies will not necessarily be devoid of life either. There are many different sorts of wetlands, supporting many different sorts of damselflies, all of which are equally "healthy" in their own right. In simpler terms, it isn't simple.

In order for damselflies to serve as indicators, it should be the case that there are wetlands out there that look, at first glance, like they are healthy—I suppose they might have plants around them, a certain water level, and ducks on the surface—but aren't. Only when you survey the damselflies—no other method can be quicker or easier, or the damselflies again lose their value as indicators—do you find out the truth. But as far as I know, this just isn't so. There are probably easier ways to measure wetland "health" than by catching and identifying damselflies, but if there aren't, at least this book will make identification easier.

The reason I'm admitting this is that I am more concerned with honesty than I am in promoting damselflies, and I'd hate to praise them for the wrong reasons. In the words of Miles Cornthwaite (a friend of a friend, who trained as a biologist and now runs a laundromat in upstate New York), we should avoid at all costs the quagmire of "pointy-headed ecology." By Cornthwaite's definition, this is simply "the delineation of the obvious by the incompetent." And hey, one of the great things about science is that a statement should, in principle, stand or fall on the basis of its own merits, and not on the basis of the status of the person who said it.

All of these considerations boil down to, at least for me, the realization that the study of damselflies is something we should do for its own sake, and not

Top: Organized odonate surveys can give interesting information on the ups and downs of damselfly populations. Bottom: This oddly purplish northern bluet might, or might not, indicate ecological trouble in the power plant canal from which it emerged.

because they are a special indicator of something else more important than themselves.

On a global scale, the intrinsic worth of odonates has recently attracted the attention of some folks in conservation circles. The people and agencies that have traditionally been concerned only with birds and mammals now see damselflies and dragonflies (and some other insect groups as well) as worthy subjects for their efforts. To date, the best summary I have seen of global odonate conservation priorities is the "Status Survey and Conservation Action Plan" prepared by the Odonata Specialist Group of the Species Survival Commission of the International Union for Conservation of Nature and Natural Resources (the IUCN/SSC). And if you think that name is long and cumbersome, welcome to the world of conservation biology. The sheer number of organizations and agencies with three- to five-letter abbreviations absolutely boggles the mind. Is this a bad thing? If it indicates a massive proliferation of ineffective, self-serving bureaucracies, I'd have to say yes. But let's get back to the topic at hand.

In the action plan document, a group of odonate conservationists have identified the major areas in which we should be concentrating in order to preserve a maximum of the world's dragonfly and damselfly species. Not surprisingly, they put the focus squarely on the tropics, while Alberta damselflies warrant no specific mentions at all. Still, the report recommends such things as habitat conservation, pesticide reduction, and encouragement of odonate research—all good ideas, if you ask me.

Of course, just because we have only 22 of the world's 2,500 or so species of damselflies doesn't mean we shouldn't care as much about them as we do about their tropical relatives. We still need to remind ourselves that most species in the province are at least locally super-abundant and seem to do well even in close proximity with boaters, cottagers, and power plants. Habitat destruction—the single biggest threat to insects today—is not yet a big enough problem to threaten damselflies here, and I can't think of any introduced creatures that have shown a tendency to out-compete, devour, or otherwise bully our native odes. All of this could change in an instant, of course, but for the moment, things in the damselfly realm in Alberta are in pretty good shape. Thus, for those who are concerned with damselfly conservation, the task at hand is not fighting for threatened or endangered animals—it is continual monitoring of large, healthy populations in case anything goes awry.

In *Tiger Beetles of Alberta*, the main conservation theme had to do with the loss of isolated patches of open, erosional habitats. The beautiful tiger beetle (*Cicindela formosa*) and the ghost tiger beetle (*C. lepida*) may not persist in Alberta if their tiny little dune habitats disappear. In the next book in this series, the one that will treat our ladybugs, the main theme will be the impact of a single, highly successful, introduced species, the seven-spot ladybug (*Coccinella septempunctata*). But with respect to the Alberta damselflies, my outlook is much less pessimistic. In fact, when we look back at the history of damselfly study here in Alberta, we see what looks like a pattern of gaining species, not losing them.

With more water on the prairies, the damselfly fauna of Alberta has burgeoned far beyond its original scope. Since 1950, Britain has lost 3 of its 42 odonate species, while in that same period of time, we here in Alberta have documented 7 species of damselflies alone that were either not present 50 years ago or were yet to be discovered in the province.

The above sentiments, more or less, are also the conclusion of a report prepared for the Alberta Natural History Inventory Centre (an agency of the provincial government) on the conservation status of the odonates of Alberta. Christine Rice, a graduate student at the University of Alberta, is the report's author. She wisely suggested that when you get right down to basics, we know almost nothing about the status of any of our damselflies, although common sense tells us there is little to worry about, at least in the short run. You might wonder why the government asked for such a report in the first place, but the fact is that many other provinces and states have "statused" their odonates, and Alberta was simply jumping on the dragon and damselfly biodiversity bandwagon.

Biodiversity is now a household word, but when I was a university student (and hey—look at my picture—I'm not an old man yet!), we never heard the term. Instead, we studied taxonomy and systematics, two names for more or less the same thing, both of which are widely considered old-fashioned today. Oddly enough, people who study the exact same thing and call it biodiversity are now considered leaders in biology.

For people like me, it doesn't matter what you call it. The documentation of diversity is still an interesting endeavour, and it is still done best by people trained in the museum tradition of biological systematics. For this reason, museums are experiencing a bit of a renaissance these days, and in order to appear progressive and innovative, many have put their biodiversity databases on-line, as "virtual museums." Those involved in government-funded conservation outside of museums have also seen the value of biodiversity databases, which they gleaned from a variety of sources including museums, naturalists, and publications.

What this means is that anyone who learns to identify a group of organisms, such as damselflies, and keeps notes or specimens documenting their findings will soon discover that their data are in demand. Not surprisingly, those individuals and institutions with the biggest databases tend to consider themselves at the hub of the biodiversity wheel, and I have watched as a number of agencies have manoeuvred and competed for the central coordinating positions in both Alberta and Canada as a whole. Frankly, I don't agree that centralized control of biodiversity data is needed. I do, however, think that such data should be readily available to those with an interest in examining it. Since collecting the

A freshly emerged vivid dancer, looking less vivid than it eventually will, perched on the boardwalk at the Cave and Basin Hotsprings.

data was a collaborative venture, examining it should be as well. This may, however, be a difficult thing to achieve, what with the territorial, competitive aspects of human nature.

The other aspect of databasing that worries me is that the form in which the data is stored can easily become outdated. When I was young, I filled out "Alberta Animal Record Cards" every time I saw a bird, mammal, fish, amphibian, reptile, or insect. It kept me busy during my teen years, I suppose. I sent them to the Federation of Alberta Naturalists, and in turn they forwarded the cards to the Provincial Museum of Alberta. There, the cards were filed in a cabinet somewhere in the back of the building. Anyone who wanted to see them could, after convincing the receptionist, have access to the cards. At the time, those of us who were keen on this project thought that every naturalist in the province would use these cards for the rest of human history. Predictably, though, they fell by the wayside and other sorts of databases took their place. In like fashion, I worry that much of what we do today will soon vanish into electronic limbo, disappear off the web, or exist only in files that are incompatible with the latest software. That is, in part, why I like writing books. By doing so, I give my ideas a much greater chance of surviving beyond the next decade or two.

As a damselfly enthusiast with an interest in conservation, I strongly suggest you do three things with the data you collect. First, store it as a "hard copy." Second, store it electronically. Third, share it with others. In these ways, you will maximize the likelihood that your observations will contribute to a better understanding of our fauna, and better conservation measures. As well, you might consider joining conservation societies that have an interest in either insects or wetlands, such as the Xerces Society and Ducks Unlimited. A very few people will actually engage in damselfly research for conservation purposes, but the rest of us can make a meaningful contribution as well.

If I had to pick a single damselfly species to be concerned about in Alberta, it would be the vivid dancer, since it lives only in the outflow of a single spring in Banff. One might think that living in a national park would be a good thing for this insect, but massive construction in and around the spring, along with numerous introductions of tropical fishes, make it clear that the habitat of this insect has been far from "protected." Today, the vivid dancers survive in good numbers, suggesting that the immense reproductive potential of damselflies makes it possible to survive all but the most permanent and severe modifications to their habitats.

History tells us that the rarest and most localized species are not necessarily the ones that disappear when people start messing with their affairs. When I was a teenage insect collector, I never would have believed that I would one day be concerned about the future of the beach tiger beetle (*Cicindela hirticollis*) or the super-common transverse ladybug (*Coccinella transversoguttata*). I am also reminded of my first summer job away from home, leading nature walks at Lac La Biche. There, in 1977, I campaigned endlessly for the lake's "endangered" American white pelicans, while dining at least twice a week on walleye and northern pike. Today, the pelicans are doing fine, while the walleye and pike numbers are collapsing. No one at the time saw it coming. No one.

None of the above says anything at all about the value of particular well-loved wetlands and their damselflies. This brings us to the reason I personally care about damselfly conservation, and that is simply the quality of life for both us and them. In a world without ponds and other wetlands, we do not encounter damselflies, and that would be a shame. Certainly there are places on earth where this exact thing has happened, and wetlands have been destroyed en masse for the sake of agriculture and other sorts of development. In my view, anything that works to preserve wetlands is a good thing. After all, anyone can get a thrill from the beauty of form, colour, and movement that damselflies provide, and wetlands themselves are tranquil, scenic places. If we forget these things, we forget something much more fundamental than "wetland health"—we forget our own mental health and well-being.

Another day, another damselfly, and a harbinger of many more to come.

7
The Damselflies of Alberta

The following treatments provide information on each of the 22 species of damselflies that are known to occur in Alberta. For each, I begin with the name, both scientific and English. I prefer to learn both, since both are now standardized. I also think it helps to have some idea of what the names mean and how most people pronounce them (most of the people I know, that is). Despite what you might hear, there is NO standard system for the pronunciation of scientific names—please believe me, or look into it for yourself if you don't. There are a number of systems for the pronunciation of Latinized words, but none has been adopted universally by biologists.

I also discuss identification of male and female adults, and how to tell newly emerged from fully mature individuals. This, combined with knowledge of which species are similar and therefore may be confused, should help you identify most of the damselflies you find. To be extra sure of your diagnoses, check the gallery of colour drawings at the end of the book, and the identification key in the appendices.

The ecology section of each species account is intended to give a sense of the habitat associations, geographic range, and flight season of each species. I have purposely chosen shaded distribution maps rather than dot maps since distributions change faster than the lifetimes of books, we know only the vague outline of each distribution to begin with.

The notes section is, not surprisingly, my favourite for each species. Here, I summarize my personal experience with each species, as well as the observations and studies of others. Hopefully, these notes will serve as the starting place for others who might wish to take the study of Alberta damselflies another step or two further down the scientific road.

When identifying damselflies there are a few other things you should probably keep in mind. First, size generally decreases with advancing season. Second, many of our pond damsels change colour with temperature. Those parts of the body that are normally bright blue or green can become dull grey overnight, only to regain their brilliance by mid-morning.

The long-legged, big-bodied river jewelwing—the supreme Alberta damselfly.

Broad-Winged Damsels

FAMILY CALOPTERYGIDAE *("Kal-OP-terr-IDGE-id-ay")*

Our single species is a member of a diverse family of damselflies that live primarily in the tropics. Most are large, as damselflies go. The most familiar broad-winged damsels do indeed have relatively broad wings, but this is not the case for all members of the family. More to the point, the wings have an almost uncountable number of tiny crossveins, indicating that they are more like those of the ancestral damselfly than are the wings of our other species, in which the venation has been reduced in the course of evolution.

Many broad-winged damsels have patterns on the wings. Some have wings that are all dark, while others have dark tips and some have dark bases. The American rubyspot (*Hetaerina americana*) has blood-red wing bases and may well show up as a wanderer in extreme southeastern Alberta (they are known from northeastern Montana). Of course, they also might never show up here, in which case I highly recommend a trip south to see them in person.

The larvae of the river jewelwing live among roots and water plants along the margins of streams. They have very broad wing pads and long slender gills.

A male river jewelwing, perched over a stream, surveying its territory.

River Jewelwing

CALOPTERYX AEQUABILIS *("Kal-OP-terr-ix ee-QUAH-bill-iss")*

There once was a angler named Wayne,
Who found fishing to be quite a pain,
So he sat by the bank,
Watched jewelwings and drank,
But he never could find them again.

IDENTIFICATION

Males: The biggest Alberta damselfly and the only one with darkened wing tips and no stigmata.

Females: Much like the males, but with obvious, white stigmata on the wings. Some do not have dark wing tips, although the entire wings may be smoky in colour.

Aging: Adults change little in appearance as they grow older—a few small areas become greyish, but not so one would notice at a distance.

Length: 43–54 mm.

Similar species: None in Alberta. Also note the broad wings that are not stalked at the base, the bright iridescent body, and the unusually long legs.

THE NAME

Calopteryx means "beautiful wing," a name that applies even more strongly to some other species in the genus, with iridescent all-black wings. "Jewelwing" means, of course, the same thing. River jewelwing is a fine name since indeed these insects prefer to live along rivers. The word *aequabilis* means "equal" or "similar" and could refer to almost anything, sort of like the old joke "What's the difference between a duck?" the punchline of which I forget (unless that was it). Dennis Paulson and Sid Dunkle have speculated that the name refers to the half-dark, half-light colour pattern on the hind wings.

ECOLOGY

Habitat: These damselflies spend all of their lives along small- to medium-sized forest streams.

Life history: Adults are on the wing in June, July, and August.

Range: Poorly known in Alberta, but probably most abundant in the eastern boreal forest. The records from west-central Alberta are discussed below.

Where to find them: The most accessible spot in Alberta is along the Beaver River near Cold Lake, where it meets Highway 881.

NOTES

In his 1918 paper, Francis Whitehouse listed the river jewelwing as a species he expected to find in Alberta. Sixty-four years later, David Maddison and I sat collecting fossils along the banks of the Blindman River near Blackfalds. As we hunched over our sorting trays, patiently searching for evidence of Paleocene life, David said something like "Hey! I didn't know we had those big, dark-winged damselflies here."

"We don't," I told him, but he insisted that one had just flown right past us. He convinced me that he knew these insects well from his childhood back east, and since David is one of the finest entomologists I know, I believed him. That became the first published record of a jewelwing from Alberta, although it later came to light that Rob Baker and Lynda Corkum had collected specimens a few years earlier in the Beaver and Medley rivers near Cold Lake. Since then, jewelwings have turned up along the North Raven River, the Clearwater River, Poplar Creek 10 km north of Fort McMurray, the Chinchaga River east of High Level, and the area north of Lake Athabasca. When Fred Korbut found the Clearwater River population northeast of Fort McMurray in 1996 (not the Clearwater River in the foothills), the species was still rare enough to warrant a note in the *Alberta Naturalist*.

Records of this species from the northeastern parts of Alberta are not surprising, since this is a common eastern damselfly that reaches its westernmost limits here in Alberta. (Note, however, that it also occurs in forested regions to the south, in extreme south-central British Columbia, and even along the U.S. west coast.) Records from the Blindman and North Raven rivers are more unusual, and may represent wandering individuals from either the east or the south. Interestingly, when Wayne Roberts found this species along the North Raven River on July 22, 1995, he found not one but three males, all together on one stretch of the creek, near the point where it meets the Raven River itself. (Wayne was fishing with his son Aaron at the time, and as fly fishermen, they—like me—prefer to call the North Raven River the "Stauffer Creek.")

Wayne spends vast amounts of time along streams in this area, and he's a very fine naturalist. When he says the jewelwings were only there once, in only one place, it's best we believe him. Probably, a single wandering female made it to the foothills, laid some eggs, and Wayne watched the resulting adults during the short time before their small population winked out of existence. It seems less likely that three males all arrived at the same spot at the same time. On the other hand, the larvae take two or three years to develop. In an area that does not support breeding populations of this species, it also seems unlikely that three larvae would survive to become adults. However, the North Raven River is fed by springs and may be able to support river jewelwings in the short run. Perhaps a hungry brown trout ate the only females to emerge that year.

These are fascinating damselflies to observe. They are only found near streams, where the males guard territories containing good patches of floating water plants, where the females can lay eggs. When a female comes along, the male changes his flight style and does a courtship display. Normally, they fly with both wings in sync, much like a butterfly. (Actually, all odonates with dark wings look like butterflies to me, at least at a distance.) The courtship display, however, involves swinging the hind wings up when the front wings are down, and vice versa, and flapping more quickly to enhance the effect. Edmund Walker once described a male displaying to a perched female, "swinging from side to side in a wide, inverted arc." Sometimes the male will flop down on the water surface, apparently demonstrating to the female how fast the creek is flowing. If these displays impress the female, the pair will mate in the usual damselfly fashion, after which she separates from the male to lay her eggs under water. He sticks around, mind you, and guards her against the intrusion of other males.

Dennis Paulson, a very fine odonatist in Washington state, recently suggested that all damselfly enthusiasts should celebrate "*Calopteryx* Week," during the first week of July, and that sometime during that week all of us should spend one day just watching the jewelwings. It's a good idea, for a thousand great reasons, not the least of which is the pleasure one gets from sitting contemplatively beside a small stream in midsummer.

A female river jewelwing, taking a break from the trials of courtship and egg-laying.

A pair of spotted spreadwings "in tandem," resting by a stream in southern Alberta.

Spreadwings

FAMILY LESTIDAE *(LESS-tidd-ay)*

This is another worldwide family of damselflies, the members of which are generally medium to large in size. Unlike the broad-winged damsels, the wings of spreadwings are narrowed at the base to form a stalk or "petiole." As well, the venation of the wings is relatively simple compared to that of broad-winged damsels, at least in the sense that there are fewer veins overall. Spreadwings can be brightly iridescent, especially when they are young, but they do not possess the brilliant blue, red, green, or yellow hues of the pond damsels.

You will find that many books on insects distinguish damselflies from dragonflies on the basis of how they hold their wings. Damselflies are supposed to hold the wings over the abdomen, while dragonflies always hold the wings out to the sides like an airplane. Spreadwings, however, break these rules and take the middle road by holding their wings out to the sides at about a 45-degree angle to the body. They also typically perch on the sides of plant stems, holding the body at a 45-degree angle as well.

Here in Alberta we have four species of spreadwings, all of which are more or less widespread and common. Thus, they do not generate the same sort of excitement as some of our other damselflies. However, there is also a fifth species that many of us believe might be hiding in the boreal forests or foothills, undetected. The lure of finding the first sweetflag spreadwing *(Lestes forcipatus)* in Alberta is enough to ensure that all spreadwings are carefully examined. Matt Holder, from Ontario, has suggested that this species may only lay eggs in sweet flag *(Acorus americanus)*, plus an introduced European species of Acorus and blue flag *(Iris versicolor)*. This may explain the sweetflag spreadwings' rarity here, since these plants are not particularly abundant in Alberta.

The sequence of adult emergence among our four species is predictable: emeralds first, then common, lyre-tipped, and finally, spotted spreadwings. This same sequence has been documented in Alberta, British Columbia, and Saskatchewan, and I use it to organize the species treatments that follow here. In British Columbia, Rob Cannings and George Doerksen suggested that by spacing themselves out through the season, the four species avoid competing with each other, both as similar-sized larvae and adults.

The egg-laying blade or ovipositor of a female emerald spreadwing is noticeably larger than that of related species.

Emerald Spreadwing

LESTES DRYAS *("LESS-teez DRY-ass")*

There once was a Lestes called dryas,
Who was cautious of cultural bias,
Alone in the swamp,
It exuded much pomp,
And appeared to be terribly pious.

IDENTIFICATION

Males: Relatively narrow pale shoulder stripe and bright green colouration. The lower claspers look like tiny inward-pointed Christmas stockings and can be seen in the field with good optics.

Females: A stocky damselfly, bright green, with a relatively long ovipositor that extends just past the tip of the abdomen.

Aging: Males become pruinose on both ends of the abdomen and the underside of the thorax, but remain obviously green. Very young males are more blue than green. Females do not become pruinose. In both sexes, the pale markings on the thorax disappear with age.

Length: Variable, averaging about 38 mm.

Similar species: Many older keys will tell you that the emerald spreadwing is the only spreadwing with a bright green thorax, but this is simply not true. Fresh adults of other species may be bright green before they begin to age. Colour is a good first clue, but always check other features as well.

THE NAME

The word *dryas* refers to the wood nymphs of Greek mythology. "Emerald" is a good way to describe these bright green damselflies.

ECOLOGY

Habitat: Found in and around ponds, often in forested or peatland areas.
Life history: This is our first spreadwing to emerge each season, and typically it appears in late June, persisting into August.
Range: Probably found throughout Alberta, except for the prairie grasslands.
Where to find them: Most of my personal encounters with this species have been in the Wagner Natural Area west of Edmonton. However, any forest pond has a good chance of supporting this species.

NOTES

Most damselfly books consider this a northern species, and perhaps its fondness for forest pools enables it to range farther into the boreal forest than most of its relatives, all of which appear to prefer warmer, more sunlit habitats. Emerald spreadwings are present all the way up to the Yukon, although they are not common that far north. At the Wagner Natural Area, I have found emerald spreadwings predictably along only one stretch of the trail, near the far end of the loop, resting on sunlit branches of black spruce and tamarack. In like fashion, Graham and Deirdre Griffiths found this species in the wooded ponds in the interior of the Clifford E. Lee Nature Sanctuary, but not in the main, open marsh.

In interior British Columbia, the Cannings brothers discovered that emerald spreadwings were unable to breed in saline lakes and that unlike the common and spotted spreadwings, they needed fresh water to survive. Typical emerald spreadwing habitats, in the Cannings' study, possessed dense submerged vegetation, cattail stems in which the damselflies laid their eggs above the water line, and a conductivity of less than 1,200 micromhos/cm. These ponds were prone to drying up during the summer. Conductivity, by the way, is not mysterious, and affordable conductivity meters (about $100) can be purchased at hydroponics stores, so that anyone can study this aspect of damselfly habitats simply by sticking the meter in the water and pressing a single button to get a readout. The conductivity of water gives a good measure of what are commonly called both salinity and hardness, regardless of which salts and minerals are involved.

This species also ranges across northern Europe and Asia, where its English name is the scarce emerald damselfly. It would, of course, be lovely if English-speaking people on all continents used the same name for the same species, but this is not likely for *Lestes dryas*. North Americans would never accept the English name "scarce emerald damselfly" since it is based on the species' relative abundance in Britain. On the other hand, it is unlikely that Europeans would use the name "emerald spreadwing" since they have another species in their fauna, the common emerald damselfly (*Lestes sponsa*), that is much more abundant and familiar. The reasons for the scarcity of the "scarce emerald damselfly" in Britain are apparently many and include draining and filling of ponds for agriculture, the use of herbicides, and habitat damage through grazing.

The iridescence on male emerald spreadwings can be truly breathtaking, before they become dusty and "pruinose" with age.

A handsome male common spreadwing, identifiable as such mainly by the shape of his claspers.

Common Spreadwing

LESTES DISJUNCTUS *("LESS-teez diss-JUNKT-us")*

There once was a wee common spreadwing,
Who bounced up and down like a bedspring,
  Adrift through the air,
  Without thoughts, dreams or cares,
'Til a frog made it into a dead thing.

IDENTIFICATION

Males: Broader pale shoulder stripes than an emerald spreadwing, with lower claspers that are shaped like a pair of tube socks (or thumbs, or tongue depressors—take your pick).

Female: So similar to female lyre-tipped spreadwings that it is best to identify them by the males they hang out with. Usually, but not always, bronze-coloured on the abdomen, with the back of the head, behind the eyes, usually black rather than yellow.

Aging: On males, the back of the head, the entire thorax, and the tip of the abdomen may become pruinose with age. On females, the pruinose areas are less extensive.

Length: Quite variable, but usually about 36 mm.

Similar species: Although this is one of our most abundant damselflies, it can also be one of the toughest to identify with absolute certainty. Males are easy to separate from lyre-tipped spreadwings. Females, however, can be green rather than bronze, and not all have black on the back of their heads (both features that are commonly used to distinguish the two). In areas where the sweetflag spreadwing might occur (the mountains are the place to look), female common spreadwings are more easily distinguished (see under the lyre-tipped spreadwing for details), while the males are almost identical. For positive identifications (under a microscope) of this tricky species pair, see *Damselflies of North America* by Minter Westfall and Michael May.

THE NAME

The word *disjunctus* means "unjoined" and apparently refers to the clearly visible segments of the abdomen, which are no more so than those of any other spreadwing—yet another indication that the first person to recognize a species is rarely the one who knows it best. The second part of the name—"common"—makes much more sense and is appropriate since this species is generally abundant.

ECOLOGY

Habitat: Common spreadwings are not terribly fussy about their habitats, except that they need emergent vegetation in which to lay eggs. They are found in ponds, lakes, marshes, slow streams, and peatland pools.

Life history: This species emerges after the emerald spreadwing, usually in early July, and is found well into September after that.

Range: Most of Alberta, except for the southeastern prairie grasslands. Also found in the Cypress Hills.

Where to find them: Almost any pond or lake at the right time of year.

NOTES

Like the emerald spreadwing, this is a species at home in the north, and in forested regions as well. The Cannings brothers show it as very common and widespread in the Yukon Territory—the only other spreadwing that far north is the emerald, which is much less common. The common is also the only spreadwing that we have found at Beaverhill Lake, east of Edmonton, a location that is too much of a lake to support lyre-tipped and spotted spreadwings, and too open and grassy to support emerald spreadwings.

The Cannings brothers found that this species has a midrange preference with respect to salinity (compared to the emerald and spotted spreadwings) and was also intermediate in its flight season, first emerging between the other two. They noted that this species requires dense submerged vegetation and that at conductivities above 1,000 micromhos/cm the only water plant that can survive is *Microphyllum spicatum*, the Eurasian water milfoil. This is an introduced waterweed with a well-deserved reputation for "choking waterways." Hopefully, it will never become established in Alberta lakes, despite its usefulness to common spreadwing larvae.

The common spreadwing also has the distinction of being the only damselfly I have ever encountered at a blacklight after dark. On a warm July night a few years back, at my favourite beaver ponds near Devon, while looking for moths, my light attracted both a male and a female common spreadwing. It is unknown whether records of this sort indicate nocturnal activity in these insects or whether I disturbed them and flushed them into the light.

Unlike most of our other damselflies, the common spreadwing has two distinct geographic races. Edmund Walker described these in 1952, naming one *Lestes disjunctus disjunctus* and the other *L. d. australis*. Some recent workers, as well as Walker himself, have suggested that the southern race (*australis* means "southern," and indeed they live to the south of the *disjunctus* race) may be a separate species. This is of little consequence here in Alberta, since the Canadian range of the *australis* race is restricted to southern Ontario, but it does influence how we think of the species as a whole.

A female lyre-tipped spreadwing. Notice the yellow splotch visible behind her right eyeball.

Lyre-tipped Spreadwing

Lestes unguiculatus *("LESS-teez un-GWICK-you-LATE-tuss")*

Those spread-winged unguiculatus,
Claimed, "All other damselflies hate us!"
   They were deemed paranoid,
   By an odonate Freud,
And relieved of their lyre-tipped status.

IDENTIFICATION

Males: The lower claspers are distinctive and big enough to see in the field. To my eye, they look like the legs of a stout little folk dancer. Once you have ruled out spotted and emerald spreadwings, and had a look at the claspers, no doubt should remain.

Females: Females are so similar to common spreadwings that I prefer to identify them by association with the males. They have more yellow on the back of the head than do the commons (the yellow extends behind the compound eyes, not just around the base of the neck) and supposedly have green rather than bronzy abdomens, although I have never found this to be consistently the case.

Aging: This species shows a similar pattern to that of the common spreadwing.
Length: About 36 mm, but variable like all spreadwings.
Similar species: If sweetflag spreadwings are ever found in Alberta, the females will be easier to recognize than the males. Female sweetflag spreadwings look like female lyre-tipped spreadwings, but with noticeably longer ovipostiors.

THE NAME

The Latinized name means "clawed grabber." The English name refers to the males' vaguely lyre-shaped claspers.

ECOLOGY

Habitat: Ponds, including temporary ones, as well as slow streams. Not a peatland species and apparently not fond of lakes.
Life history: Similar to other spreadwings, overwintering as eggs. In Alberta, adults emerge in July in most places and persist into late September.
Range: Found in the prairies, parklands, and at lower elevations in the mountains.
Where to find them: Prairie sloughs later in the summer will predictably produce this species.

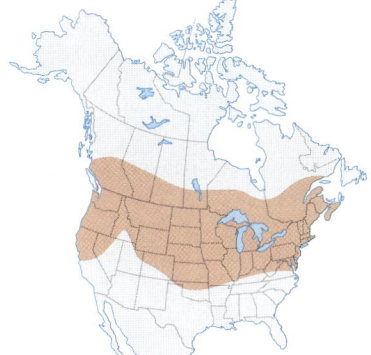

NOTES

The books I learned from state that female lyre-tipped spreadwings have green abdomens while female common spreadwings have bronze abdomens. This, unfortunately, doesn't seem to hold here in Alberta. For example, some of my early field notes from the Wagner Natural Area read "saw many females with lots of yellow on back of head: some were entirely bronzy, others were bronze at base and green near the abdominal apex. Hmph." As well, at Chappice Lake, on the prairies north of Medicine Hat, all of the female lyre-tipped spreadwings I have seen are bronzy, not green, in colour. Presumably, our Alberta spreadwings show different colouration than those living farther to the east.

Bill Sawchyn and Cedric Gillott studied this species in Saskatchewan and found that its life history was similar to that of the emerald and common spreadwings. They found that all three species overwintered as eggs in late stages of development (in contrast to the spotted spreadwing, which develops little before the snow flies). This is the typical pattern for spreadwings in temperate areas, and their eggs can survive at up to −20 °C. This is a common winter temperature around Saskatoon, so the eggs need adequate snow cover to prevent death from extreme cold and desiccation. Snow is an excellent insulator, and a few decimetres can easily prevent the temperature from falling below −20 °C at ground level. Hatching occurs at about 10 °C, about a month after the ponds fill with rainwater, near the end of the first week in May. The larvae develop rapidly and emerge as adults after about two months. The adults all emerge in about a 10-day period and reach sexual maturity in 16 to 18 days. Immediately after copulation, females lay eggs in green bulrush stems, 5–60 cm above the water line. They select stems in small groups or on the periphery of large stands, since these are more likely to be broken in winter and covered with snow.

A male lyre-tipped spreadwing, in the typical spread-winged pose of these damselflies.

Blue-eyed and beautiful, a male spotted spreadwing stops briefly at a flower, with no interest at all in pollen or nectar.

Spotted Spreadwing

LESTES CONGENER ("LESS-teez con-JEN-ner")

There once were some Lestes congener,
Who lived in a slough east of Jenner,
Each summer it dried up,
Some larvae were fried up,
So meadowlarks ate them for dinner.

IDENTIFICATION

Males: This is our easiest spreadwing to identify, thanks to two dark spots on the side of the thorax behind the base of the rear leg (as opposed to an all-pale or all-dark area). Males have short lower claspers that look like shoeprints.

Females: Both sexes are relatively dark overall and have narrow pale shoulder stripes.

Aging: As they get older, spotted spreadwings become darker until they are almost black. The tops of the eyes become deep blue, and the usual patches of pruinosity appear, more so on the males.

Length: About 34 mm, but variable.

Similar species: Usually smaller and darker than other spreadwings.

THE NAME

Lestes means "robber" or "grabber" and refers to the way these damselflies (or perhaps their larvae) capture their prey. *Congener* means "of the same sort" and probably reflects the similarity of this species with other types of spreadwings. The spots in "spotted spreadwing" are the dark spots on the sides of the thorax.

ECOLOGY

Habitat: Seems to prefer ponds to lakes and will also use slow streams as larval habitat. Often, their breeding ponds dry up by late summer. Rarely found in peatlands.

Life history: Spotted spreadwings have a typical spreadwing life history, but are the last species to emerge as adults, usually sometime in July. Where they occur, they are usually the last damselflies on the wing in the fall.

Range: Primarily a prairie and parkland species, rare farther north than Opal (about an hour's drive northeast of Edmonton).

Where to find them: Almost any prairie slough in late summer.

NOTES

In the early 1970s, Bill Sawchyn and Cedric Gillott studied the life history of this species near Saskatoon, and at the time they wrote that "in spite of its distribution throughout temperate North America…*Lestes congener* Hagen is probably the least studied member of the Odonata of this region." One of their most interesting findings was that, unlike other spreadwings, the eggs of spotted spreadwings overwinter in a very early stage of development. For some reason, this makes them more resistant to cold than the eggs of the other species, and they can survive $-33\,°C$ and extreme desiccation without difficulty. They start developing as soon as they are wetted, even at close to $0\,°C$. At the same time, the eggs of their spreadwing relatives have already developed to an advanced stage, and the spotted spreadwing eggs develop quickly in order to catch up—something they almost but don't quite achieve.

Spotted spreadwing eggs hatch in late May and on into June, and the larvae develop quickly, in about 50 days. Once the adults emerge, sexual maturity takes three weeks to achieve. The females lay eggs in dry bulrush stems (although other odonatists have seen them lay eggs in a variety of plants, both living and dead, including willows, cattails, and rushes), always in tandem with males, and an average female can lay 205 eggs in her lifetime. They survive into October, even when the temperature dips to −4.5°C (but not past −9.5 °C). Sawchyn and Gillott mention that all of these features are adaptations to seasonal ponds at high latitudes, but caution that the life history of the species as far south as the Mexican border in California should be examined as well, which so far, to my knowledge, it has not been.

At Riske Creek in British Columbia, the three Cannings brothers found spotted spreadwings breeding abundantly in lakes that were too salty (greater than 5,000 micromhos/cm) for other spreadwings, with a pH above 9. Here, their only odonate neighbours were boreal bluets. The Cannings noted that spotted spreadwings emerged as adults about 10 days after the common spreadwings and 20 days after the emerald spreadwings; the three species taking turns using the lake.

Since spotted spreadwings emerge late in the season, their larvae often live in lakes with increased salinity due to evaporative concentration of salts. As salinity increases (to a maximum of 13,214 micromhos/cm), the larvae develop more quickly, perhaps to outrun the rising salt levels. In highly saline lakes, they are tightly associated with Juncus balticus, the wire rush. Without the plant, no larvae were present. However, spotted spreadwings also breed readily in freshwater lakes with almost no salinity at all, and it is best to think of them as salt tolerant, not salt-adapted.

The spots of this female spotted spreadwing are peeking out at you from her flanks (see also page 58).

A study in blue, green, black, and brass: the female taiga bluet.

Pond Damsels

FAMILY COENAGRIONIDAE *(See-NAG-ree-ON-idd-day)*

The pond damsels are our most diverse family of damselflies, and they are worldwide in distribution. In general, they are medium to small in size and quite colourful. In contrast with the spreadwings and jewelwings, their colours go beyond mere iridescence and pruinosity. Pond damsels are pigmented, for the most part, in bright hues of blue, green, red, orange, or yellow. For most people, these are the typical damselflies of Alberta.

The wings of pond damsels are stalked and are usually held together over the abdomen. Only one species in our fauna is habitually known to perch with its wings outspread: the western forktail. Some pond damsels are fond of perching in vegetation, while others are often seen resting on the ground.

The first damselflies of spring are invariably pond damsels, although some species have later flight seasons than the others and persist into September. As far as we know, all of the Alberta members of this family spend the winter as larvae. Pond damsel larvae are much like those of our other two families, but have gills that are more rounded and leaflike, in that the tracheal breathing tubes leave the main trunk at an acute angle (not a right angle) and branch into many finer tubules. They also have a shorter lower lip than spreadwings or jewelwing larvae.

Eurasian Bluets

GENUS COENAGRION (see-NAG-ree-on)

Bluets are generally blue and black in colour—at least the males are—but not all blue and black damselflies are bluets. The official bluets are divided into two groups: the Eurasian bluets (genus *Coenagrion*) and the American bluets (genus *Enallagma*), named for their respective centres of diversity. Most Eurasian bluets live in Europe and north Asia, but we have all three of the North American species here in Alberta. On the other hand, most American bluets are found in North America. Adults of the two groups are relatively easy to tell apart, but the larvae are extremely similar. Most books will tell you that Eurasian bluet larvae have seven antennal segments while American bluets have six. When Ed Fuller showed me specimens with six on one antenna and seven on the other, I quit believing that. Adult male Eurasian bluets have green on the underside of the thorax, or a broken pale shoulder stripe (that looks like an exclamation point), or both, whereas male American bluets are all blue and black (here in Alberta, that is). Adult female Eurasian bluets do not have a spine on the underside of the eighth abdominal segment, while those of American bluets do.

A young male taiga bluet rests on a marsh marigold—an early-season damsel on an early-season flower.

With its green-bottomed eyes, the male prairie bluet is easy to recognize, and easy to admire.

Prairie Bluet

COENAGRION ANGULATUM *("See-NAG-ree-on ang-you-LATE-um")*

> *There once was a C. angulatum,*
> *Who grabbed up mosquitoes and ate 'em,*
>   *They never bit her,*
>   *So it didn't occur,*
> *To that damsel that we people hate 'em.*

IDENTIFICATION

Males: A distinctive pattern on the long segments of the abdomen: each is almost entirely dark, with the thin blue bands becoming progressively shorter toward the abdominal tip, which is bright blue. The top of the second segment has a moustache-man face on it, or at least a moustache.

Females: Dark abdomen like a taiga bluet, but with a pale base on the first of the short segments near the tip (the eighth segment, if you can see to count them). Usually yellow-green to brownish, but sometimes blue like the males.

Length: About 30 mm.

Similar species: Relatively easy to recognize. See above for separating females from those of taiga bluets.

The Damselflies of Alberta

THE NAME

The "coen" in *Coenagrion* means "shared." This name was coined to replace *Agrion,* the former name for these insects, with which it "shared" the same meaning. The word *angulatum* means "angled" or "cornered," and probably refers to the shape of the male claspers. This species is not restricted to the prairies in Alberta, but it does indeed occur there, so the English name is quite appropriate.

ECOLOGY

Habitat: Found in open, sunlit ponds, sloughs, marshes, and slow streams.

Life history: Bill Sawchyn and Cedrick Gillott studied this species near Saskatoon, where adults emerge in late May (when the average daily water temperature is about 12°C and the maximum air temperature reaches 20°C), then mature, and lay eggs that hatch after 16 days and grow to become almost full-sized larvae before mid-October.

Range: Most abundant on the prairies, but also found throughout the aspen parklands in appropriate habitats and in the southern boreal forests in open, sunlit areas.

Where to find them: This is a species that you should run across early in your damselfly-chasing career, as it is both abundant and widespread.

NOTES

The prairie bluet is an easy species to recognize by its colour pattern alone, but that doesn't mean that it is impossible to mistake it for something else. In the older keys, a wing vein feature is used to separate the bluets from the forktails, among other things. It just so happens, however, that some male prairie bluets have the same venation as forktails, and this has led to any number of mistaken identifications of these damselflies as eastern forktails. See under that species for more details.

In British Columbia, this species is known only from the Peace River grasslands in the northeast. This population is probably also found on the Alberta side of the border and is probably separated geographically from other populations of this species to the south.

Bill Sawchyn and Cedric Gillott's study of this species deserves further comment. They found that the maximum adult emergence occurred about ten days after the first emergence and that the adults mature in about ten days, away from the ponds in which they developed, eating mainly midges. They lay eggs under the water line, and the pair can crawl down emergent plant stems and remain under water laying eggs for half an hour, at a depth of up to 45 cm. Females lay an average of about 172 eggs.

When the larvae stop growing in the fall, they purposely place themselves where they will become embedded in ice. Some cling to plant stems, while others crawl upside down onto the underside of the ice as it is forming and let it freeze around them. Generally, they place themselves 15–20 cm below the ice surface. Amazingly, the larvae themselves do not freeze, since they possess antifreeze compounds in their blood, at least during the winter. This nifty adaptation may protect the larvae from predators that might otherwise eat them during the winter, but it is more likely a way of utilizing shallow habitats that predictably freeze—an example of going with the flow, even if the flow freezes solid.

It is possible that this habit of freezing into the ice might be one factor that limits the geographic range of the prairie bluet. After all, larvae that overwinter in liquid water cannot become any colder than 0°C, for obvious reasons, whereas those that embed themselves in ice can become considerably colder. In the boreal forest, predictably heavy snow cover insulates the ice and prevents it from becoming as cold as it might on a bare, windswept prairie slough at an air temperature of −25°C or less. G. R. Daborn, a master's student at the University of Alberta and the first person to find Eurasian bluet larvae frozen in ice, observed higher winter mortality rates than did Sawchyn and Gillott. In response, the latter two authors suggested that lack of snow cover may have been the cause of these differences.

Of our three species of Eurasian bluets, this is the odd one out. The other two are more closely related to each other than either is to the prairie bluet, and this shows in their choice of habitats. The prairie bluet is not a peatland species, whereas the other two often are. As well, if you look at them carefully, you'll see that the other two are also built more alike and that the prairie bluet is the most heavy-bodied of the three. All three of our species have closer relatives in the Old World than they have among each other.

More colourful than the male, this female taiga bluet is one of our most charming common damselflies. For the male, see page xii, or page 7.

Taiga Bluet

COENAGRION RESOLUTUM *("See-NAG-ree-on rez-oh-LUTE-um")*

There once was a resolute bluet,
Who just couldn't wait to get to it,
  He flew past a bridge,
  And snatched up a midge,
Then sat on a thistle to chew it.

IDENTIFICATION

Males: A distinctive abdominal banding pattern on the long segments: two short, dark bands near the thorax and then a single long one formed from the dark parts of two and a half segments (I look for the "two-and-a-half band" in the field). The top of the second segment has a demon-head pattern on it. Mature males also have a lovely blue colour on the top of the body and a lighter green on the underside.

Females: Abdomen almost entirely dark, with very narrow, light blue bands only on the short segments nearest the tip. Can be either yellow-green or blue-grey in ground colour.

Length: About 30 mm.

Similar species: Female marsh and Hagen's bluets also have all-dark abdomens, but both of these species also possess a spine on the underside of the eighth abdominal segment. Male taiga bluets with broken pale shoulder stripes look like subarctic bluets, but subarctic bluets never have the two-and-a-half band pattern on the abdomen.

THE NAME

The word *resolutum* means "resolute," which in turn means "firm of purpose." In other words, this species tolerates a northern existence (though we all know that it isn't so bad!). Taiga is another name (a Russian word) for the boreal forest.

ECOLOGY

Habitat: Abundant in woodland ponds, weedy streams, marshes, grassy ditches, and spring-fed pools. Whereas the prairie bluet seems to prefer sunlit habitats, the taiga bluet is more often found in waters that are at least partly shaded.

Life history: Adults first emerge in late May most years. In the mountains, I have seen them on the wing as late as September 5 (at the Vermilion Lakes), but in the lowlands they are usually gone by the end of July.

Range: Probably found throughout Alberta, but not terribly common on the prairies.

Where to find them: Another widespread, abundant species that should be hard to miss.

NOTES

Taiga bluets are undoubtedly my favourite damselflies. I've known them all my life, I've seen them in remarkably huge numbers at their breeding ponds, as well as here and there all through the city of Edmonton, and I always get pleasure from their lovely colours and fascinating habits. Along with two dragonflies proper—the boreal whiteface (*Leucorrhinia borealis*) and the hudsonian whiteface (*L. hudsonica*)—they are the first odes of the year for me, most of the time.

Taiga bluets probably breed farther north than any other species of damselfly, well up into Alaska and the territories (boreal bluets have been found farther north than taiga bluets, but not in breeding populations). They also frequent the northern tier of the United States, where they bring a touch of the "far north" to our American odester friends.

On some male taiga bluets, the pale shoulder stripe is broken to form an exclamation mark, much like that of the subarctic bluet. This is also the case in British Columbia, and Edmund Walker noticed that in Ontario the farther north one goes, the heavier the black markings are on this species and the more likely the pale shoulder stripe will be broken. It may be that someone will look at geographic variation in this species and give subspecies names to two or more races, but for the moment we do not have sufficient information to do anything of the kind.

Near Saskatoon, Bill Sawchyn and Cedrick Gillott found that this species has a life history much like the prairie bluet. However, taiga bluets stop growing in early October as larvae (they stop growing earlier than larvae of prairie bluets), and the adults are not known to lay eggs under water. Another interesting aspect of this species' biology is that mating begins away from the ponds, not at them.

In another Saskatchewan study, M. Forbes, B. Leung, and G. Schalk studied fluctuating asymmetry in this species. "Fluctuating asymmetry" refers to the fact that some individuals are more symmetrical than others—the right side is a more perfect mirror image of the left, and vice versa. This is thought to be the result of better genetic control of growth, which in turn might indicate a better set of genes in general. The theory is that more symmetrical individuals should show better survival than their asymmetrical pond-mates and that females should choose more symmetrical males with which to mate. Having said this, the three Saskatchewaners found, in essence, nothing. But in the long run, they might well be vindicated. Increasingly, biologists are coming to realize that the very faint statistical "signals" that they are trying to measure in fluctuating asymmetry studies are often lost in the "noise" of measurement error and other sorts of variation. This has led to a large number of false positive results, based more on wishful thinking and statistical accidents than on real biology. One of the most clear-headed commentators on fluctuating asymmetry has been Rich Palmer, a zoologist at the University of Alberta, and many of his ideas can be found on his personal website:
http://www.biology.ualberta.ca/palmer.hp/palmer.html.

The exclamation mark on the top of its thorax is an almost-sure sign that you are looking at a subarctic bluet!

Subarctic Bluet

COENAGRION INTERROGATUM *("See-NAG-ree-on in-terr-oh-GATE-um")*

There once was a bluet subarctic,
Who believed in that ol' Noah's Ark schtick,
When she heard me insist,
It was all just a myth,
It made her poor damselfly heart sick.

IDENTIFICATION

Males: Black markings on the underside of the thorax and a broken pale shoulder stripe (also found on some *C. resolutum*) are distinctive. The male abdomen has a "two-and-a-bit" blue section at the tip, rather than the typical two. The top of the second abdominal segment has a Pokemon face on it (and if this book outlives Pokemon, think "big-eyed cute face" instead).

Females: Thin, light bands on abdomen; usually blue like the males, although some are greenish.

Length: About 30 mm.

Similar species: Males can be confused only with male taiga bluets with broken pale shoulder stripes, but male subarctic bluets never have a dark "two-and-a-half band" on the abdomen. Females are distinctive.

The Damselflies of Alberta

THE NAME

The word *interrogatum* means "questioned," but nowadays no one knows what question was on Hermann Hagen's mind when he named the species in 1876. Hagen was used to writing his manuscripts in Latin and presumably felt no need to explain himself further. "Subarctic" is an overstatement here in Alberta, but this species is indeed the most northerly of its group.

ECOLOGY

Habitat: Found in cool water, typically beaver ponds or pools in peatlands with aquatic mosses in the water and sedges around the edges.

Life history: Probably similar to that of the taiga bluet, with adult records from Alberta ranging from late May through August. In the Yukon, Rob and Syd Cannings saw some evidence that subarctic bluets emerge sooner than taiga bluets in the spring.

Range: A species of the boreal forests and the mountains, but not found in the Cypress Hills.

Where to find them: There is a beaver pond about 25 km south of Highway 16 along Highway 753 (north of Cynthia) that has produced subarctic bluets for many years now.

NOTES

Many decades ago, Edmund Walker called this "our least known *Coenagrion*," and this is still the case today. The larva of this species was completely unknown until Rob and Syd Cannings found and described it in 1980, as did Rob Baker and Hugh Clifford. (This coincidental discovery resulted in both scientific papers being published back-to-back in the same journal, since they contained slightly different information on the same topic.) Even where it occurs, the subarctic bluet is generally less common than the taiga bluet, which is found in almost all places where you find subarctic bluets, although the reverse is not generally true. When Bill Sawchyn and Cedric Gillott made their study of the taiga and prairie bluets near Saskatoon, there were subarctic bluets in the same pond. Unfortunately, the subarctic bluets were not included in the study, since they were uncommon. This leaves a great opportunity for a modern Alberta odester to study and document the life history of this species. I have no such plans as it stands, so please feel free to beat me to it.

Is this a more northerly species than the taiga bluet? Possibly, since its southern limit is farther north than that of any other damselfly in North America. Will it move even farther north if the climate becomes warmer? Possibly, but the distribution of this species in Alberta needs to be better documented before we can say.

Subarctic bluets are typically localized and rare, and live in low-density populations. Here are two anecdotes from my field notes that illustrate how elusive this species can be. First, when Natasha Page did her study of the odonates of the Wagner Natural Area in 1995, she found subarctic bluets only in the east beaver pond, and even then only a few. The following year, that pond dried up and we have not seen the species at Wagner since. Second, Su-Ling Goh and I found subarctic bluets in small numbers at a dugout pond near the beaver pond mentioned above, along Highway 753 in 1996. When I returned in 1998, I found that the area around the pond had been burned in a forest fire and the species was no longer present, or at least was no longer detectable. I have not been back since.

A beaver pond near Cynthia—great habitat for the rare subarctic bluet.

American Bluets

GENUS ENALLAGMA *(ENN-nah-LAG-mah)*

When most people in Alberta think of a damselfly, they think of an American bluet. This is our most diverse damselfly genus, with seven species in the province. Not only that, it is also our most difficult genus, at least with respect to identification. In many instances, identifying an American bluet is impossible without having the damselfly in your hand. To identify American bluets in the field, first assess their relative size. Two (Hagen's and marsh bluets) are noticeably smaller than four of the other five, and this is obvious in the field. With males of all species, look for the relative size of the postocular spots and the relative lengths of the first three dark bands on the abdomen (on segments 3 to 5). With females, look at the extent of light markings near the tip of the abdomen (especially on segment 8) and the exact shape of these markings.

Why are the American bluets all so similar? It is because they have diverged from their common ancestor very recently. This is the conclusion of a bright and innovative researcher named Mark McPeek. Mark is originally from Kentucky and is now a professor and department head at Dartmouth College in

A crab spider devouring an American bluet.

New Hampshire. He has worked with a number of other biologists on his bluet research, most notably Jonathan Brown at Grinnell College in Iowa and Michael May at Rutgers University in New Jersey. In McPeek's opinion, many of the American bluet species originated mere moments ago, at least by the usual standards of evolutionary thinking. By analyzing DNA samples from various bluet species, Mark has estimated that many of the bluet lineages split off from their nearest relatives quite late in the Ice Ages, 50,000 years ago or less. To me, this was a whopping big surprise. My background in beetle studies taught me that the insect species that we see around us now have remained more or less the same for at least a million or two years. Fossil beetles from peat deposited during the Pleistocene epoch (the last 1.8 million years) are generally indistinguishable from living insects. Another yardstick for comparison is our own species, Homo sapiens, which has been around for about 250,000 years. Friends of mine who do DNA studies with other sorts of insects have also expressed skepticism about McPeek's findings, but so far his research seems to be holding up to scrutiny. Thus, we have here something more than a group of beautiful creatures that are tough to identify—we also have a cutting-edge debate about the speed of evolutionary change in insects.

Another of Mark McPeek's contributions has been based on the observation (by Dan Johnson and Phil Crowley) that these damselflies fall into two subgroups with respect to their habitats, at least in eastern North America. The first are "fish lake species" that live in places where fishes are the top predators (mainly bass-family fishes—*Centrarchidae*—including smaller things such as sunfish). The second are "dragonfly lake species" in which large darner dragonflies (family *Aeshnidae*) are the top predators. In the Alberta context, in the absence of bass-family fishes, the equivalent predators might be perch and pike, although so far it seems that this may not be the case. Certainly, trout-family fishes (trout, char, and whitefish) are ineffective predators on damselfly larvae, due to their cruising feeding style in still waters, at least when compared to bass and sunfish. According to Mark McPeek, minnow-family fishes and sticklebacks are also unimportant predators of damselfly larvae, as are giant diving beetles *(Dytiscus spp.)* and giant water bugs *(Lethocerus americanus)*. It seems unlikely that Alberta bluets select their habitats in this fashion, but it also appears reasonable to believe McPeek's assertion that this distinction was important in the early evolution of the group. Two of our species, the boreal and northern bluets, evolved in dragonfly lakes, while the rest are fish lake species.

A typical male boreal bluet, as blue as blue can be.

Boreal Bluet

ENALLAGMA BOREALE *("ENN-nah-LAG-mah BOR-ay-AHL-ee")*

There once was a bluet so boreal,
Living north of the land of the oriole,
* It was often her wish,*
* To avoid all the fish,*
So it said on her pondside memorial.

IDENTIFICATION

Males: One of the larger bluets. Males are a typical blue and black damselfly with three short dark bands on the long abdominal segments, then two long ones. The first three are similar in length and relatively short, at least in most places (the short-and-equal pattern). Some populations have longer bands that increase in length toward the end of the abdomen.

Females: Overall ground colour ranges from light blue to yellow-green. The abdomen is mostly dark, with significant light areas on segments 8, 9, and 10; the front margin of the black mark on the top of abdominal segment 8 tapers to a thin line (i.e., the sides of this point are concave) or the segment may be all blue. The rear margins of the shoulder pads are incomplete.

Length: About 32 mm.

Similar species: Male northern bluets are virtually indistinguishable from male boreals, except for the shape of the upper claspers. Females are also extremely similar, except for the rear margin of the shoulder pads and sometimes the form of the dark mark on abdominal segment 8.

THE NAME

Enallagma is a difficult name to interpret, but apparently comes from a Greek word meaning "crosswise." Of course, there are many things on an *Enallagma* (body markings, wing veins, etc.) that are indeed "crosswise." The word *boreale* means "of the north," hence the English name as well.

ECOLOGY

Habitat: Ponds and lakes, generally without fishes in them. Also common in peatland pools. Can tolerate relatively alkaline, salty water.

Life history: Adults emerge in late May or early June, but are generally gone by mid-August.

Range: Widespread and common.

Where to find them: This is an easy species to find, just about anywhere in Alberta.

NOTES

Rob Cannings and Kathleen Stuart, in their book on the odonates of British Columbia, comment on the fact that boreal and northern bluets rarely occur in the same places, but that no one knows what separates them. I have found this to be the case in most Alberta localities as well. Typically, when I first arrive at a site, I catch a few of the boreal/northern bluets and examine them in the hand with a lens. If I find a mix in the first 20 specimens, then I examine each individual I find—otherwise I assume that they are all the same.

Most of the male boreal bluets in Alberta have the typical look that characterizes the species, with three short, dark bands on the basal half of the abdomen. In the cooler habitats of the foothills, mountains, and northern forests, these bands are longer and the bluets look more like some of our other species. Edmund Walker wrote that this species was more slender in the south and more robust in the north. He saw specimens "with the dark areas greatly increased" from northern Manitoba, the Mackenzie District, the Northwest Territories, and the Yukon Territory, "giving them the appearance of a different species." In his description of these individuals, he points out that some have divided pale shoulder stripes, like those of subarctic bluets.

Since this colour pattern is found only in a particular geographic area, it probably qualifies for official recognition as a subspecies. In fact, if I had enough specimens I would probably do that myself, and I may still get around to it. But why has no one else beat me to the task? Probably because damselfly scientists distrust the notion of subspecies and because this one lives in the far north, a long way from the heart of North American academia.

Since I've spent a great deal of time studying both tiger beetles and butterflies, I am much more comfortable with the subspecies concept. As well (and this goes without saying, really), I'm always eager for a chance to promote the scientific value of Alberta insects. So, for the moment, I will refer to the dark boreal bluets as the "boreal/montane race" and hope that the idea catches on. As an added complication, some workers feel that the boreal bluet is already a subspecies, of the European *Enallagma deserti*. Mark McPeek, however, has found that DNA evidence suggests that these two species are actually quite distantly related.

A pair of boreal bluets in the heart position. Not all females are this yellow in colour.

A rather attractive female northern bluet, with an interesting mix of yellows, blues and browns in its colour pattern. And don't mistake the dashes on the sides of its abdomen for those of the vivid dancer!

Northern Bluet

ENALLAGMA CYATHIGERUM *("ENN-nah-LAG-mah SYE-ath-IDGE-er-um")*

There once was a real cyathigerum,
Who was truly a very good midger, some...
   thought he should eat
   More mosquito meat,
But he stuck to his diet in oblivium.

IDENTIFICATION

Males: One of the larger bluets. The male is a blue and black damselfly with a short-and-equal pattern to the first three dark abdominal bands, and relatively large postocular spots.

Females: Overall ground colour light blue or yellow-green. Abdominal segments 9 and 10 are pale on top, and sometimes 8 as well, although many females have a triangular mark on the hind margin of this segment.

Length: About 32 mm.

Similar species: Both sexes are almost identical to the boreal bluet. Males are best distinguished by the form of their claspers, while females are best identified by the complete rear margin of the shoulder pads, or the triangular front margin of the dark mark on the top of abdominal segment 8 (as opposed to a concave triangle, when this mark is present).

THE NAME

The word *cyathigerum* means "cup-bearer" and refers to the shape of the black marking on the top of the male abdomen. "Northern bluet" is a good name for this species, although it does occur as far south as Baja California.

ECOLOGY

Habitat: Ponds, lakes, and marshes without perch and pike, but rarely alongside the boreal bluet. Unlike the boreal bluet, this is not typically a peatland species.

Life history: Emerges a bit later than the boreal bluet, typically in early June, and disappears by mid-August in most places, although I do have one record from the very end of August at Oster Lake in Elk Island National Park.

Range: Widespread and common in Alberta.

Where to find them: An easy species to find.

NOTES

Northern bluets are found all the way to about 70 degrees north in Siberia and 68 degrees in the Yukon, making them some of the most northern damselflies known (along with boreal bluets and a Eurasian bluet, *Coenagrion concinnum*). Since the northern bluet was the first in its genus to be described and named, it is the "type species" of the genus and is also one of only two American bluets found in Europe. The genus as a whole is found on all continents except Antarctica and Australia, although some feel that the African species should not be considered true *Enallagma*.

This species can exist in extremely large populations. I've encountered huge numbers myself, and Dennis Paulson tells me that he saw so many at a small lake south of Lac La Biche that there was a blue cast to the scene. (Of course, lakes are often blue, but you get the point.) In Europe, northern bluets have been shown to tolerate pollution exceptionally well, although adults in polluted waters are often smaller than average. They can live in water with a conductivity of 8,000 micromhos/cm (that is really salty) and a pH of 3–4 (which is to say, very acidic).

The mating habits of this species have been well-studied. Ola Finke showed that males are important in rescuing females when the latter emerge from underwater egg-laying and get caught in the surface tension. The males attach their claspers to the females' necks and tow them to safety without getting caught in the water themselves. I saw exactly this phenomenon when we taped the odonate episode of *Acorn, The Nature Nut*, and the resulting videotape is a highlight of that program.

In the north, Edmund Walker recognized a darker form of this species, much like the boreal/montane race of the boreal bluet, but did not give it a subspecies name. Males have increasing bands on the abdomen, rather than the short-and-equal pattern. Walker saw dark specimens from Terrace, British Columbia, and Admiralty Island, Alaska. I have seen such individuals at McGregor Lake (east of Milo) and at Taylorville, but these are not particularly northern localities and so far I'm not convinced that we see the same pattern in both boreal and northern bluets in Alberta. Westfall and May say that some have divided pale shoulder stripes, but I have not seen Alberta specimens with this feature either.

The post-ocular spots of male northern and boreal bluets are larger than those of their relatives.

To something as small as a male Hagen's bluet, the tip of a thistle leaf is actually a comfortable place to rest.

Hagen's Bluet

ENALLAGMA HAGENI ("ENN-nah-LAG-mah HAW-gen-eye")

There once was a bluet of Hagen's,
Who knew not a thing of toboggans,
  "Winter is nice,
    Frozen tight in the ice"
Was the thought that rolled round in its noggin.

IDENTIFICATION

Males: One of the smaller bluets. On males, the first three abdominal dark bands increase in length toward the tip of the abdomen. In some places, this species can be separated from the marsh bluet only on the basis of the claspers. In others, the two may be distinguishable at a distance (see below for details).

Females: Ground colour usually yellow-green or light brown, sometimes blue, sometimes blue on the abdomen and greenish on the thorax. Abdomen almost entirely dark. Otherwise indistinguishable from marsh bluets except for the raised hind margins of the shoulder pads.

Length: About 30 mm.

Similar species: Marsh bluets are similar in size and in colour pattern. Females can also be confused with female taiga bluets, but the latter do not have a spine on the underside of abdominal segment 8.

THE NAME

This damselfly was named in honour of Hermann August Hagen, a Prussian physician who wrote much of his most important work on North American damselflies before he had ever set foot on our continent. He spent the last 26 years of his life at Harvard. Hagen's bluet was named for him in 1862, five years before he left Europe, by B. D. Walsh, an amateur American entomologist who also translated one of Hagen's important manuscripts from French to English.

ECOLOGY

Habitat: Ponds, marshy lake margins, slow streams. Occasionally found in peatland pools.
Life history: Adults emerge from June to August, later than the boreal bluet.
Range: Widespread but so far not known from the mountains, or north of the latitude of Lesser Slave Lake.
Where to find them: This is one of our common species, and damselfly enthusiasts should have no trouble finding them early in their careers.

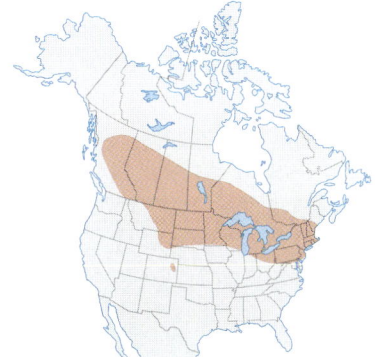

NOTES

Separating this species from the marsh bluet is an excellent way to prove to yourself the value of local knowledge. At Big Chickakoo Lake and Wabamun Lake, for example, male Hagen's bluets are larger, and have the second and third dark bands shorter than do male marsh bluets. At the Vermilion Lakes, the Hagen's bluets are small and dark, like marsh bluets at the other localities. Yet in the standard books, the two are said to be indistinguishable except for their claspers.

    The mating habits of this species have been well studied by Ola Finke, and in fact this is the best-known member of its family in this respect. Of special interest, apparently, is the fact that males and females find one another at the "hinterland rendezvous," a very romantic-sounding place if you ask me. (Actually, it just means "away from water" in this context.) All of our other damselflies (except for vivid dancers and eastern forktails) meet their mates at the place where they plan to lay eggs. Finke was also able to determine the time between egg-laying bouts for an individual female (about six days)—a bit of information that is lacking for all of our other species. This means that females have six days during which they can either find a better mate for their next batch of eggs or avoid being accosted by unwanted suitors. In other words, she can choose to either visit or stay clear of the hinterland rendezvous, where males wait on lookout perches for passing females.

Francis Whitehouse, in 1918, predicted that this species would be found in Alberta, but had no records of it himself. If we assume that it was as common then as it is now, it is unimaginable that he could have missed it. This suggests that Hagen's bluets (along with marsh bluets, alkali bluets, and tule bluets) were part of an American bluet invasion that has taken place in Alberta over the last century or so.

The heavily-marked female of Hagen's bluet, looking noticeably old and worn-out while resting on a fern.

A male marsh bluet, at Buffalo Lake, whose first three dark abdominal bands quite typically increase in length from front to back.

Marsh Bluet

ENALLAGMA EBRIUM *("ENN-nah-LAG-mah EE-bree-um")*

There once was a damsel in a marsh,
Whose life was exceptionally harsh,
Her language was clean,
A real odonate queen,
But when push came to shove she'd say, "Garsh!"

IDENTIFICATION

Males: Another smallish bluet, much like Hagen's, and much like a prairie bluet, but bluer. Males with first three dark bands increasing noticeably in length, and long to begin with, making this the darkest of our American bluets.

Females: One of two species with very little in the way of light abdominal markings. Ground colour blue or, less often, yellow-green.

Length: About 30 mm.

Similar species: Most male Hagen's bluets have shorter abdominal bands than marsh bluets, but some (for example, in Banff) are as dark or darker. Fortunately, the claspers are a dead giveaway. Females only separable from Hagen's bluets on the basis of the non-raised margin of the shoulder pads.

The Damselflies of Alberta

THE NAME

The word *ebrium* means "drunk" or "sated." This species was named by Hermann Hagen, and it has been suggested that Hagen thought it looked inebriated in flight. However, since Hagen had never seen one alive when he named it (he had yet to set foot in North America), one has to wonder what was really on his mind. "Marsh bluet," by comparison, is a very sensible name indeed.

ECOLOGY

Habitat: A pond and lake species (sometimes found in slow streams) that apparently does not do well in peatlands and is at home alongside yellow perch and northern pike.

Life history: Another species that first emerges in June and finishes its adult season sometime in August.

Range: Probably absent from the mountains and the extreme north, but widespread elsewhere in Alberta.

Where to find them: No special instructions are needed—you'll find them easily if you look.

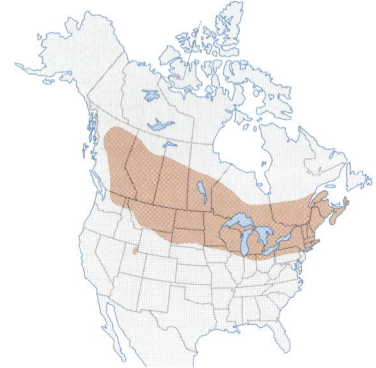

NOTES

Like Hagen's bluet, this is a species that was expected but unknown when Francis Whitehouse wrote his 1918 paper on the damselflies and dragonflies of Alberta. That Whitehouse could have missed it is impossible, unless it was much less common in his day. This seems likely, and I now believe that we have seen three waves of American bluets come into Alberta. The first consisted of northern and boreal bluets, the most cold-hardy of the lot. Then came Hagen's and marsh bluets, which were known by the time Walker published in the 1950s. And finally the tule and alkali bluets arrived, to be detected in the 1990s. This is, of course, just a guess, but it makes more sense to me than the assumption that our predecessors were incompetent in the field!

Mark McPeek found that this species does best in lakes that are occasionally "winter-killed" such that only a few fish survive. The bluets benefit from the presence of a limited number of fish because their larvae are easily killed by large dragonfly larvae, and the fish keep the dragonfly numbers down. Too many fish, on the other hand, can be a significant threat in itself. It is difficult to say if this is the case in Alberta or not, where winterkill is a function of lake size and depth (bigger deeper lakes are less prone to oxygen deprivation) on the one hand, and snow cover and long ice-covered periods (snow-covered lakes and long winters make winter-kill more likely) on the other. McPeek also showed that adult marsh bluets are prone to leave their home waters in search of better habitats—lakes at just the right stage of winter-kill recovery—but how they recognize such places is anyone's guess.

As with all damselflies, egg laying is a big strain on the females. Mark Forbes and Rob Baker have shown that the number of eggs a female can lay goes down with increasing levels of parasitism by water mite larvae (the little red dots that you often see clustered around a damselfly's leg bases). On the other hand, it's clear that the mites have ample opportunity to give up on the damselfly and return to the water, since J.-G. Pilon, in Quebec, timed one female laying eggs under water for a total of five hours without coming to the surface for air!

Marsh bluets mating, with a very male-like bright blue female in back.

A good-looking male alkali bluet from Gull Lake, Alberta.

Alkali Bluet

ENALLAGMA CLAUSUM *("ENN-nah-LAG-mah KLAW-zum")*

One day on the ol' bald-butt prairie,
Where damselflies find it real scary,
There once was a bluet,
Who up and said, "Screw it!
I'm leaving the Bleriot Ferry."

IDENTIFICATION

Males: Variable in length, but generally like a slightly darker version of a boreal/northern bluet. In males, the first three abdominal bands are roughly equal and of medium length, postocular spots are small and not connected by a central bar, and claspers are distinctive.

Females: With a pit on the pronotum, and the abdominal segment 8 entirely pale. Ground colour blue or yellow-green.

Length: 24–37 mm.

Similar species: Males are the only medium-and-roughly-equal species in our fauna, and females are most like those female boreal/northern bluets with a blue segment 8. Both sexes are easiest to identify when they are smaller than boreal/northern bluets.

THE NAME

The word *clausum* means "a closed space" (like a closet). It apparently refers to the shape of the claspers (the usual explanation in entomology when nothing else makes sense, the person who coined the name didn't explain it, and is also now dead). "Alkali bluet" refers to the habitat of this species, which is often slightly saline lakes.

ECOLOGY

Habitat: A lake species that does well in moderately saline waters but is not restricted to them.
Life history: A long, late flight season, according to Walker, probably meaning mid- to late June through August.
Range: Lakes in the southern parkland and prairie regions.
Where to find them: Gull Lake, Interlakes Wetland immediately north of Brooks, Prairie Oasis park north of the Sheerness power plant.

NOTES

I was tickled to find this species for the first time in Alberta during the summer of 1995. My aunt has a cabin in the summer village of Gull Lake, and I've been chasing insects there since I was a child. In fact, I published a number of damselfly records from Gull Lake back in 1983. So finding alkali bluets in good numbers was a surprise, since I doubt I would have missed them 12 years earlier. As well, although I have no proof that he ever visited Gull Lake, it is worth noting that Francis Whitehouse was based in Red Deer and collected extensively in the Red Deer area. Gull Lake is now only a 20-minute drive from Red Deer and was always a popular place to visit, although never as popular as Sylvan Lake (which to my knowledge has no alkali bluets).

If you know Gull Lake, you know that it is indeed an alkaline place. The beach has a salty crust on it most of the time, and the water is shallow. This is exactly the sort of place that alkali bluets like, and Gull Lake is certainly not alone in providing this sort of habitat.

This female river bluet is still young and is therefore not as brightly coloured as she will eventually become.

River Bluet

ENALLAGMA ANNA *("ENN-nah-LAG-mah ANN-nah")*

There once was a bluet named anna,
"Exclusively Americana,"
  But she hid near the creeks,
  For decades, not weeks,
'Til they found her just north of Montana.

IDENTIFICATION

Males: A relatively large bluet, found near prairie streams. In males, first three dark bands (on segments 3–5) increase in length with the shortest near the base and the longest near the tip of the abdomen. The claspers are large enough to see at a distance on this species, as a pair of long, slightly upturned pincers.

Females: Much like boreal, northern, or tule bluets, but usually with thin dark shoulder stripes. Segment 8 can be almost all dark or almost all pale (that is, blue or yellow-green).

Length: About 32 mm.

Similar species: Males are much like male alkali bluets, but perhaps with thinner dark shoulder stripes. Females are best told from boreal, northern and tule bluets by the form of the shoulder pads.

THE NAME

Anna Tribolet was the future wife of E. B. Williamson when he named this species. The English name is easier to interpret, and indeed this is our only *Enallagma* that prefers streams to ponds and lakes.

ECOLOGY

Habitat: Near slow prairie streams, but apparently not found along irrigation canals.
Life history: All of our records are from June and July.
Range: Ross Creek in Medicine Hat (near Strathcona Island Park) and Fish Creek in Calgary.
Where to find them: Both locations are open to the public, and the damselflies are relatively easy to find near the creeks.

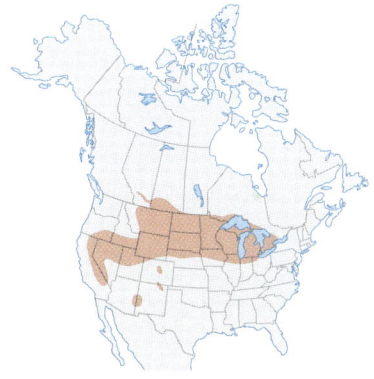

NOTES

In Edmund Walker's three-volume treatise, there were only four American bluet species in Alberta: the boreal, northern, Hagen's, and marsh. The river bluet became the fifth species in 1980, when Dan Soluk (an Albertan now working as an entomologist in Illinois) found them in the Ross Creek in Medicine Hat. This was also the first record for this species in Canada. They were also found in Fish Creek in Calgary, and to my knowledge these are still the only two locations known for this species. Even where it occurs, it is not common, and in my opinion it could easily have avoided detection in Walker's time.

River bluet larvae live in streams only, and in this respect they are unique among our bluets, since all the others are more at home in the still waters of ponds and lakes. This may also explain why they are less common than the other bluets, since stream-dwelling damselflies rarely reach the same population densities as those in ponds and lakes.

Although I've seen river bluets at Fish Creek, most of my experience with this species has been in Medicine Hat, near Ross Creek. Luckily, males are easy to identify in the field, since their claspers are large enough to be recognizable at a distance.

In 1995, the South Saskatchewan River flooded, and Ross Creek became part of a sea of brown water that enveloped the entire neighbourhood through which it flows, along with much of Strathcona Island Park. My wife's parents, who lived along the river below Redcliff, would have had water 2.5 metres up the sides of their house had they not sandbagged and built an earth dam between themselves and the river. I was, of course, concerned for my in-laws, but I was also worried about the river bluets. The next season, however, with the creek back to its usual level (and plenty of dried muck hanging in the trees to remind us of what happened), the river bluets were back, as if nothing had happened. Apart from some stubborn skidloader tracks on their lawn, the same could be said for my wife's parents.

This species seems especially prone to hybridizing with others, oddly enough given its great whopping claspers. Hybrids have been reported elsewhere in its range with both tule bluets and familiar bluets (*E. civile*—another species that may well turn up in Alberta sooner or later). In fact, Mark McPeek's DNA studies have shown that the river and tule bluet are more closely related to each other than they are to the rest of the Alberta members of the genus *Enallagma*.

The whopping big claspers on this male river bluet are obvious, even at a distance.

A male tule bluet, resting among dried plant stems on the margin of the Interlakes Wetland, near Brooks.

Tule Bluet

ENALLAGMA CARUNCULATUM *("ENN-nah-LAG-mah kar-unk-you-LAY-tum")*

There once was a bluet, no fool,
Who asked is it "tu-lee" or "tool,"
You really can't blame her,
Laugh, point at, or shame her,
For that bluet had not been to school.

IDENTIFICATION

Males: A bigger bluet from the prairies. Males, like a larger Hagen's or marsh bluet, with increasing dark bands on the abdomen and distinctive claspers.

Females: Segment 8 partly pale at the base and segments 1–3 darker than in boreal/northern bluets; postocular spots smaller. Ground colour blue or yellow-green.

Length: About 32 mm.

Similar species: The combination of large size and increasing length of the dark bands on segments 3–5 make males most like river bluets, but without the great whopping claspers. Females are most like boreal/northern bluets, but see above.

The Damselflies of Alberta   103

THE NAME

The word *carunculatum* means "a small piece of flesh," a name that would seem appropriate for almost any insect, but actually refers to—you guessed it— the shape of the male's sexual claspers. "Tule" refers either to bulrushes (plants in the genus *Scirpus*) or the marshy areas in which bulrushes live. The word, by the way, is pronounced "TU-lee."

ECOLOGY

Habitat: Prairie lakes and reservoirs, sometimes in moderately salty water.
Life history: Appears later in June than the early bluets and persists through August and into early September.
Range: Exclusively a prairie species in Alberta.
Where to find them: I recommend the Interlakes Wetland, just north of Brooks. There is a parking lot there, and a short trail leading to an observation platform. Along the sides of this trail, tule bluets (as well as alkali, northern, boreal, and Hagen's bluets) are common.

NOTES

For some time, the only tule bluet known from Alberta was a single specimen that I caught at the Tilley Reservoir near Brooks. Then Christine Rice and Jon Hornung did their field study in the Brooks area and turned up a number of different locations, and many dense populations. Thus, our most recently discovered American bluet went from being a "rare" element in our damselfly fauna to being an expected species on the prairies.

Like the alkali bluet, I suspect that it was not here all along. Since the time of Francis Whitehouse, and especially after the Dust Bowl era, the amount of water on the prairies of Alberta has changed dramatically. Irrigation, cattle watering dugouts, waterfowl management areas, power plants, and recreational areas have all provided habitat for damselflies (and other aquatic creatures) that simply was not present until relatively recently. Thus, I think of the tule bluet as a recent immigrant to Alberta, making use of the new habitats that we have generously provided.

Forktails

GENUS ISCHNURA *(ish-NEW-rah)*

Compared to the bluets, forktails are relatively rare in Alberta. When Edmund Walker wrote *The Odonata of Canada and Alaska*, there was only one species known from the province, and it was known here from only one lake. Now we have records for four of the six Canadian species, and it is time to take this genus seriously.

Forktails are well known for their long flight seasons and the two colour morphs that most species show among the females. In general, it seems that orange or tan-coloured females are less likely to be killed by predators, while blue or green male-like females are subjected to fewer mating hassles, from both their own species' males and those of other species. Not all studies support this explanation, mind you.

Forktails reproduction is also unusual because the adults mature much faster than most other damselflies (sometimes in a single day) and the females of some species (in Alberta, at least the eastern forktail) apparently mate only once. In most forktail species the females lay eggs alone, without the benefit of their mates guarding and protecting them from other males.

Although our four species belong to a single genus, they are clearly divisible into two pairs of similar species. Male Pacific and plains forktails both have the pale shoulder stripes reduced to four dots, while male eastern and western forktails have full pale shoulder stripes, green thoraces, and blue-tipped abdomens. Females, unfortunately, are more alike than the males. The combination of a green or orange thorax and a blue-tipped abdomen makes many of our forktails the only Alberta damselflies with two contrasting colours, and they are for this reason some of my favourites to watch and to photograph.

The Pacific forktail—for many decades the only known member of its group in Alberta.

With dull colours and pale pterostigmata, it is clear that this female Pacific forktail is something out of the ordinary.

Pacific Forktail

ISCHNURA CERVULA *("Ish-NEW-rah SIR-view-lah")*

There once was a fork-tailed cervula,
Who practised the miniature hula,
 With her long abdomen,
 She could sometimes twirl ten,
And create quite a hullabaloola.

IDENTIFICATION

Males: A blue and black damselfly with the pale shoulder stripe reduced to two spots (giving a four-spotted thorax pattern), and a dark abdomen with a blue tip.

Females: Most have an orange-brown thorax without dots or pale shoulder stripes, and a pale tip to the abdomen. Some lack the pale tip, and male-like females are rare.

Aging: Greyish pruinescence can obscure the markings of older females.

Length: About 30 mm.

Similar species: Male plains forktails have a lower "fork" on the "tail" (the top of segment 10). Female plains forktails have nipple-like bumps on the top of the prothorax, as well as tufts of hair, and lack a protuberance on the rear of the prothorax.

THE NAME

Ischnura means "slender tail" and is a fine description of the abdomen of these insects. The word *cervula* means "deer" (specifically, the wapiti and its relatives) and probably refers to the forked projection at the tip of the male abdomen (which is a poor imitation of an antler, if you ask me). This species is found along the western side of the continent, so "Pacific forktail" is an appropriate English name.

ECOLOGY

Habitat: Outside Alberta this species is found in a wide range of ponds and slow streams, and is common in slightly salty, alkaline waters. In Alberta, it is found only in a spring-fed lake and a few beaver ponds.

Life history: Adults are probably on the wing from late May through September. The length of the larval stage in Alberta is not known, and in the cool waters of the Vermilion Lakes it may exceed a year.

Range: In Alberta, the spring-fed Vermilion Lakes in Banff National Park and a few beaver ponds in the Cypress Hills.

Where to find them: Pacific forktails are especially common along the margins of the Third Vermilion Lake in Banff.

NOTES

The Pacific forktail was another one of Edmund Walker's interesting discoveries on his historic trip to Banff, during which he found vivid dancers at the hot springs and the first notopteran "rock crawlers" known to science. Compared to the latter two discoveries, finding a damselfly more characteristic of British Columbia on the eastern side of the continental divide must have seemed like a fairly minor thing. It is interesting, though, that this species has not spread down the windy Bow Valley and out onto the prairies, where its usual sort of habitat is present in many places. Instead, one finds the plains forktail on the prairies, albeit sparingly, and even then only recently. In Alberta, the Pacific forktail behaves like a mountain species.

The population of Pacific forktails that Don Hilton found in the Cypress Hills is therefore most likely a true relict, having spread there during a cooler climatic period (shortly after the last glaciers melted, some 10,000 years ago), after which all of the intervening populations became extinct. On the other hand, it also seems likely that the Pacific and plains forktails are probably the evolutionary result of the division of a formerly widespread species into eastern and western populations. It is conceivable that these may now be meeting again for the first time, and the possibility of finding intermediate, hybrid forktails should not be ignored.

The male Pacific forktail is one of only two species of damselflies in Alberta with a pattern of four blue dots on a black upper thorax.

Can you beat the beauty of this red, pink, and blue female plains forktail?

Plains Forktail

ISCHNURA DAMULA *("Ish-NEW-rah DAM-mue-lah")*

Those damselflies out on the plains,
Will forever be pleased when it rains,
If it doesn't they'll seek,
Out new ponds, lakes, and creeks,
And with luck pioneer new domains.

IDENTIFICATION

Males: Like the Pacific forktail, a blue and black damselfly with the pale shoulder stripe reduced to two spots, and a dark abdomen with a blue tip.

Females: Most have an orange-brown thorax without dots or pale shoulder stripes, and a pale tip to the abdomen. Some lack the pale tip, and male-like females are rare.

Aging: Older females do not develop very much grey pruinescence, and it rarely obscures their colour pattern.

Length: About 30 mm.

Similar species: Compared to Pacific forktails, male plains forktails have a much lower "fork" on the tenth abdominal segment, while on females the prothorax has nipple-like bumps on the top and lacks a blunt knob in the middle, facing backward.

The Damselflies of Alberta 109

THE NAME

The word *damula* means "small deer" and refers to the even more pitiful antler-like projection on the males, compared to that of the Pacific forktail. Plains forktails, not surprisingly, live mainly on the plains, but also some distance away, in springs.

ECOLOGY

Habitat: Lakes, ponds, and power plant outflows in Alberta.
Life history: Probably a single year per generation and a long flight season for the adults.
Range: A variety of ponds and lakes in the eastern half of Alberta.
Where to find them: Near Edmonton, the power plants at Wabamun Lake provide the most impressive populations, while damselfly enthusiasts in the south should visit Strathcona Island Park in Medicine Hat.

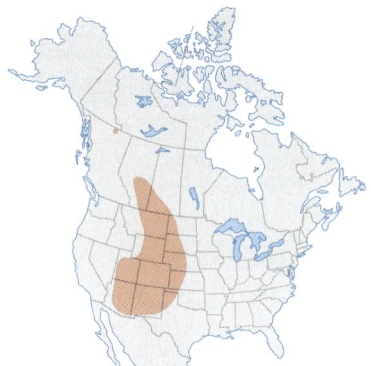

NOTES

The history of this species in Alberta is interesting. In July 1995, Natasha Page began finding what she thought were forktails during her survey of the odonates of the Wagner Natural Area. I cautioned her that they were probably prairie bluets with unusual wing veins and that this had tricked many people in the past. Most of them turned out to be just that, but while we were taping a television show at Wagner, Natasha produced a male that was clearly a four-spotted type of forktail. We photographed it and videotaped it, but then it suddenly flew up and away, leaving us not knowing if it was a Pacific or a plains forktail. A few weeks later, on the first of August, Carole Patterson and I came upon huge numbers of Pacific forktails in the warm water outlet canals of the Wabamun town power plant (and later found that they are also present in the outflows of the Sundance and Keephills plants). This seemed to solve the mystery: the plains forktail is a species that is found in hot springs in British Columbia, so when it somehow found its way to the power plants at Wabamun, it treated them as artificial springs. Wagner, being downwind of

Wabamun most of the time, probably picks up a few stragglers blown away from the main population. Or so we thought.

Records then began to appear from places without springs. I found them in Medicine Hat, Carroll Perkins photographed them at Hastings Lake, Ed Fuller caught a male at Beaverhill Lake and another at Ironwood Lake, I found more at Pigeon Lake and near Brooks, and Christine Rice and Jon Hornung found them at three more locations on the prairies. I now think of this species as a recent immigrant to Alberta (surely the early collectors could not have missed it!) and a damselfly that is at home in a variety of typical Alberta damselfly habitats.

In Alberta, the Pacific forktail is a spring-associated species and the plains forktail is more widespread. In British Columbia, the opposite is true, and the plains forktail is found only in the Liard River Hot Springs in the northern part of the province. That population attracted a great deal of attention in British Columbia and is now on the "red list" in that province, meaning it is considered in danger of localized extinction. Rob Cannings regards the Liard River Hot Springs population as a relict from a widespread distribution during the warm period between 5,000 and 6,000 years ago, about half the time since the last glaciers melted away. Thus we may have a warm-period relict forktail (the plains) in northern British Columbia and a cool-period relict forktail (the Pacific) in the Cypress Hills of southern Alberta—an odd sort of symmetry.

Sadly, in 1981, a 41-year-old odonatist named George Doerksen was killed by a grizzly bear while camping at the Liard River Hot Springs and studying these very damselflies. At the time, he was working on a project much like this one, photographing and documenting all the odonate species in British Columbia.

A male plains forktail, photographed soon after the plains forktails were discovered in the warm water outflow of the Wabamun Power Plant.

A male western forktail from the distinctly western town of Medicine Hat.

Western Forktail

ISCHNURA PERPARVA ("Ish-NEW-rah per-PAR-vah")

There once was a forktail so western,
That even the varmints quit pesterin',
*   It ate up the skeeters,*
*   That the rustlers and cheaters,*
Might otherwise find plum molesterin'.

IDENTIFICATION

Males: Green markings on the thorax, a blue-tipped abdomen, and complete pale shoulder stripes.

Females: When freshly emerged, females have an orange-tan thorax, and a dark abdomen with an orangish base and either a pale or a dark tip. A male-like form is not known for this species.

Aging: Females become entirely blue-grey and pruinose with age, at which point all other colours and patterns are obscured.

Length: About 25 mm.

Similar species: Male eastern forktails have a pointed tip on the lower clasper (it is C-shaped in the western forktail), and female eastern forktails have lower shoulder pads than westerns and never have a pale-tipped abdomen.

Western forktails typically hold the wings closed at rest, while eastern forktails sometimes hold them slightly open, like a spreadwing.

THE NAME

The name *perparva* means "very small," which it is. This species, the eastern forktail, and the sedge sprite are our smallest damselflies.

ECOLOGY

Habitat: Elsewhere, found in ponds, often alkaline, with a mud substrate. This more or less matches the habitat in Medicine Hat.
Life history: Unknown in Alberta, but probably with a long flight season and a single year per generation.
Range: So far, known only from Medicine Hat.
Where to find them: The ponds at Strathcona Island Park in Medicine Hat are the most accessible Alberta location for this species.

NOTES

The trail that leads out from Strathcona Island Park in Medicine Hat and along Ross Creek is one of my favourite damselfly spots. There, the river bluet used to be the most sought-after species and the only species that was unusual in the area. Things changed in the summer of 1998 when I discovered large populations of both plains and western forktails in the Strathcona Island Park ponds. So far, this is the only recorded location for the western forktail in the province.

When I mentioned my discovery to local naturalists, they told me that it was widely known that odonate larvae had been deliberately placed in the ponds at Strathcona Island Park for mosquito control and that Medicine Hatters were proud that they hadn't resorted to pesticides. I began to hear stories of trucks full of swamp water being brought up from Montana and dumped in the city ponds. This struck me as unlikely, and I asked my friends to look into this further. As it turned out, the story was entirely untrue. Thus, it seems that all of our Alberta damselflies got here of their own volition, including the western forktail, which is also known from southern British Columbia, and Manitoba, but so far not Saskatchewan.

An eastern forktail male, photographed near Sudbury, Ontario, but identical to Albertan members of this species.

Eastern Forktail

ISCHNURA VERTICALIS *("Ish-NEW-rah vurr-tick-AL-iss")*

There once was an eastern Ischnura,
It must be! I checked and I'm sure—
   A trained person takes pains,
   To examine the veins,
But if the key's wrong, then so's my bravura.

IDENTIFICATION

Males: Very much like the western forktail, with a green thorax, complete pale shoulder stripes, and a blue-tipped abdomen.

Females: Again, very much like western forktails, usually orange-tan in colour with a dark abdomen. Male-like females are rare (although no females of any sort have been found in Alberta so far).

Aging: Females become grey and pruinose with age, but not as much as western forktails.

Length: About 25 mm long.

Similar species: Male western forktails have a C-shaped tip on the lower claspers, while females have higher shoulder pads and always have a dark-tipped abdomen. As well, eastern forktails are apparently more prone to holding the wings partly open at rest, like a spreadwing.

THE NAME

Verticalis refers to the vertex, which is the top of the head, and the name was given to this species by the great American entomologist Thomas Say. Unfortunately, Thomas didn't say what he meant at the time, so we don't know what aspect of this species' vertex he considered notable. It is, however, black.

ECOLOGY

Habitat: A pond species, but not so tightly associated with mud-bottomed ponds.

Life history: As with our other forktails, this species probably has a long flight season and takes a single year per generation.

Range: Known from a few locations in the Brooks and Lake Newell area.

Where to find them: The most accessible location is the Kinbrook Marsh near the entrance to Kinbrook Island Provincial Park, south of Brooks. There, Christine Rice tells me that the forktails are easiest to find low among the rushes and cattails to the west of the floating dock.

NOTES

In almost all books on damselfly identification, wing vein features are used to separate forktails from other pond damsels. In forktails, the fourth and fifth branches of the radial vein (the third major vein from the leading edge of the wing) diverge near the fourth crossvein past the nodus on the front wing, and near the third on the hind wing. In the bluets, this fork is farther out on the wing.

Now to be fair, Edmund Walker wrote in the *Odonata of Canada and Alaska* that "this character is useful but not invariable," but this warning was far too brief and understated to prevent tremendous confusion down the line. In 1983, I published a bogus record of the eastern forktail from George Lake, near Busby, Alberta, and was mortified to discover my error many years later. It was a prairie bluet instead, with wing veins of the forktail type. These same sorts of prairie bluets confused Natasha Page at the Wagner Natural Area, and I also found an entire specimen tray filled with "*Ischnura*" in the collections of the Provincial Museum of Alberta, all of which were prairie bluets. In the E. H. Strickland

Entomology Museum at the University of Alberta, there is a male taiga bluet (not a prairie) that E. H. Strickland himself identified as the pacific forktail (the label reads "*Ischnura cervula* Teste EHS," meaning that E. H. Strickland himself attested to its identity). By this point, it was widely "known" that the eastern forktail was found in Alberta, despite nothing but misidentified specimens. What a relief it was, then, to see three males collected near Rolling Hills by Christine Rice and Jon Hornung on June 18, 1999. I have examined them carefully, and this time we have the genuine article for sure. Since then, Christine and Jon have found additional localities in the Brooks area.

To my mind, the oddest thing about this history of confusion is that the four species involved (eastern and Pacific forktails, as well as prairie and taiga bluets) are pretty easy to recognize, even at a distance. Sure, some preserved specimens lose quite a bit of their colour pattern, but still, there is usually enough left to help make a quick, reliable identification (which is how the mistakes were eventually spotted). I suspect that what was to blame here was scientific fustiness of an old-fashioned sort. Colour patterns were long considered lesser features for insect identification, whereas things like wing veins and genitalia are considered much more objective and concrete. The same was, of course, true for birds—early ornithologists were inclined toward physical measurements of bill length and such, and proving that birds could be identified by colour patterns in the field was a hard-won triumph for the Roger Tory Peterson crowd.

This is a well-known and well-studied damselfly. Ola Finke, a prominent odonatist, found that the females mate only once and that males recognize mature females by their pruinosity. Eastern forktails are also known for their ferocity—they are the feisty little weasels of the damselfly world. George Bick and Lothar Hornuff reported seeing eastern forktails chase off much larger lyre-tipped spreadwings on numerous occasions, and Philip Corbet cites a record of a female killing and eating a freshly emerged spreadwing. In Quebec, Don Hilton saw one eat a sprite, but compared to bullying and murdering spreadwings, this seems like a minor accomplishment.

Singular Damsels

The last three species to be treated here are all the sole members of their genera in Alberta. As such, they give hints at the diversity of damselflies farther afield. The dancers (*Argia*) are a large genus with 36 species in Canada and the United States, and even more in Latin America. Our single species, the vivid dancer, is only able to live this far north because it specializes in breeding in warm springs. In contrast, the red damsels (*Amphiagrion*) are a small genus, with only two species. One lives in eastern North America, and the other is western. Like the vivid dancer, the western red damsel is associated with springs in Alberta, but in this case cool springs will do, and warm springs are apparently too warm for its liking. Our third singular damsel is a sprite (*Nehalennia*), and the sprites are found in both North America and Europe, with six species overall. It is unlikely that additional species in any of these three groups will be found in Alberta in the future, but more surprising things have certainly happened.

The sedge sprite—a tiny, iridescent damselfly with no post-ocular spots.

A mature male vivid dancer, perched on a fallen tree trunk, in Banff National Park.

Vivid Dancer

ARGIA VIVIDA *("Ar-JEE-ah VIH-vih-dah")*

There once was a young vivid dancer,
As mean as a Florida panther,
*   He drank, and he smoked,*
*   Didn't care—he just joked,*
'Cause his life was too short to get cancer.

IDENTIFICATION

Males: Blue and black, with the front portions of the long abdominal segments light on top and dashed with black on the sides. Some mature males are purplish, and at low temperatures, the colours of these damselflies darken. This is an almost robotic damselfly that sometimes walks while perched and often simultaneously raises the wings and abdomen in a quick, jerky movement that looks like a nervous deep breath.

Females: Even splotchier than the males. There are two female colour morphs. One morph is blue and black (the "andromorph," meaning male-form), while the other is orange-brown and black (the "heteromorph," meaning other-form).

Aging: Newly emerged adults are milky off-white and black before becoming blue and black.

Length: About 32 mm.

Similar species: Roughly resembling some American bluets, but with a very different abdominal pattern (and bluets are never whitish).

THE NAME

Argia means "lazy," but it was probably intended to mean "bright" (from the Greek *argos*), while *vivida* refers to this animal's vivid colours. "Vivid dancer" is a fine name for this damselfly.

ECOLOGY

Habitat: In Alberta, this species breeds only in the warm sulphurous pools of the Cave and Basin Hot Springs. Adults perch in the nearby woods, on the ground, on logs, and on the railings along the boardwalk around the springs.

Life history: Overwinters as larvae, and emerges as adults from June through August, persisting into September. In Alberta, they have a one-year life cycle.

Range: Restricted to the Cave and Basin Hot Springs.

Where to find them: If I were you, I'd try the Cave and Basin Hot Springs.

NOTES

Edmund Walker was the first to find this species in Alberta, and wrote that "this Sonoran species is abundant at Banff…the distribution suggests a connection with the hot springs there." Now, we know that just about every hot spring in the west, up to the latitude of Banff, has vivid dancers living in it, but no other localities are known from Alberta. According to Gordon Pritchard, who has made the vivid dancer a major research subject, it may be that climate has prevented vivid dancers from ranging farther north, or perhaps they haven't had time to get there yet. There are none in places such as the Liard Hot Springs in northern British Columbia, for example. It is interesting, as well, that vivid dancers have been found in at least eight cool springs in the southern Okanagan in British Columbia, by Rob Cannings and his colleagues.

At Halcyon Hotsprings in British Columbia, Kelvin Conrad and Gordon Pritchard showed that male vivid dancers control other males' access to females but not to resources such as egg-laying sites. The males wait on sunlit places on the forest floor, about 10–15 metres from the water, and there they intercept females to mate with. They then mate for about 40 minutes and remain in tandem for two to three hours until the time is right for egg laying, during the middle of the day. Other males come to the water in the afternoon, find females that for some reason have no males, and mate with them for 10 minutes or so before going straight to egg-laying sites. Kelvin and Gordon suggested that this is a widespread general pattern for damselflies, although their study was the first to show it.

They also studied the two colour phases (dark and bright) induced by temperatures above and below 20°C, and the two colour morphs of females (blue and red-brown). Once the adults are warm enough to exhibit the bright phase, they also reflect ultraviolet light quite strongly, probably making it easier for them to see each other at a distance. However, males showed no preference with respect to blue versus red-brown females. It is possible that blue females avoid harassment by males, since they look like males themselves. This has been shown elsewhere in the Rambur's forktail (*Ischnura ramburii*). It is also possible that there is a tradeoff between being easy to find (blue) and being cryptically protected from predation (red-brown). Conrad and Pritchard were unable to resolve this question. In vivid dancers, the two morphs are about equally common, but in other species the red-brown females usually outnumber the blue ones.

Gordon Pritchard also showed that the life cycle of the vivid dancer at Banff is synchronized and takes one year. Since the water is warm enough for the larvae to develop throughout the year, this synchronization is important—otherwise, adults could emerge in mid-winter or full-grown larvae could sit around for months waiting for the right conditions to transform to the adult stage. It turns out that short day-length causes older (but not full-grown) larvae to enter an inactive (diapause) phase of their lives, in which they do not grow. When the days begin to lengthen in the spring, the larvae can grow again, and this ensures that they emerge as adults in June, July, or August.

Pritchard has also identified 12 species of odonates at the Cave and Basin Hot Springs, but only vivid dancers and four-spotted skimmers (a dragonfly proper, *Libellula quadrimaculata*) are found as larvae in the springs, and only the dancers can complete their life cycle in the warm water. The water, by the way, is an almost constant 26°C year-round.

The Cave and Basin Hot Springs have been in continual ecological upheaval since the park was first formed. The most obvious disturbance came from the fact that they were channeled through a great pool, inside a spa building. They were also once home to a unique population of long-nosed dace minnows (*Rhinichthys cataractae smithi*), a fish that is now apparently extinct. Aquarists in Banff intentionally released any number of tropical fishes into the spring waters, and now all you see there are mosquitofish (*Gambusia affinis*), sailfin mollies (*Poecilia latipinna*), and jewel cichlids (*Hemichromis bimaculatus*). For years, there were rumours that Parks Canada was going to poison these introduced fishes, but this never came to pass. It may be that they were afraid that a few dace might have survived undetected, and they didn't want to kill off the last ones by accident.

Gordon Pritchard's work in Banff is a good example of how difficult it has been (there are signs that things are getting better lately) to study insects in the national parks. In Pritchard's words, it amounted to "years of frustrations," during which parks officials "just rejoiced in saying no" each time he proposed another research project. At one point, Gordon and his students were even turned down when they wanted to do a mosquito survey in Banff. They were allowed to collect mosquitoes in the Vermilion Lakes, but not in the Banff townsite itself. Ironically, the mosquitofish in the Cave and Basin Hot Springs were introduced in 1924, by park managers, in an attempt to eradicate these same mosquitoes. Somehow, even today, many parks managers honestly seem to feel that people like entomologists are a threat to be guarded against, while roads, hotels, ski slopes, towns, garbage dumps, and all manner of construction projects are somehow innocent and allowable.

Things for Gordon Pritchard took an odd turn when Parks Canada tore up the springs to make a new boardwalk in an operation that involved stream diversion, bulldozers, and disturbances far greater than an army of odonatologists could have caused in a lifetime. When officials at Parks Canada realized that they might have destroyed the northernmost population of vivid dancers, and that Pritchard might blow the whistle on them, they began treating him more graciously and granted him two permits: one for his own vivid dancer studies and one for delegates at the 1983 International Symposium of Odonatology to collect specimens in the park that summer. The records collected on that permit are still of great interest today, especially to people like me, and of course none of the odonate populations in Banff were affected by the activities of the odonatists. Eventually, Pritchard decided that it was far easier to drive a bit farther and work outside the national park boundaries.

The male western red damsel is like no other damselfly in Alberta.

Western Red Damsel

AMPHIAGRION ABBREVIATUM *("AM-fee-AG-ree-on ah-BREE-vee-AT-um")*

There once was a damsel so red,
Who woke up one morning and said,
I love each little thing,
About this little spring,
So I'll stay here until I am dead.

IDENTIFICATION

Males: A red and black damselfly, and our stockiest and hairiest damselfly as well. In general, this is our only orange or red and black damselfly without postocular spots.

Females: Orange and black, and generally lighter in colour than the males.

Aging: Dark areas expand with age in both sexes.

Length: About 27 mm.

Similar species: All other reddish or orange female damselflies have postocular spots, and usually some blue and green body markings as well. No other species in Alberta has males that could be confused with western red damsels.

THE NAME

Amphiagrion means something like "around the field" and refers to the grassy places the adults hang around in. The word *abbreviatum* means "shortened," referring to the relatively short body of these insects, compared to other damselflies. "Red damsel" is self-explanatory, and there are two species in North America, the other one being the eastern red damsel, *A. saucium*.

ECOLOGY

Habitat: Locally abundant in spring-fed ponds.
Life history: Whitehouse said they emerge during the second week of June.
Range: Widespread, but localized in spring-fed pools. The range of this species in Alberta is not yet well defined.
Where to find them: Any of the locations mentioned in the text below.

NOTES

Edmund Walker was the first to find this species in Alberta, and wrote of finding a mating pair in "a small open marsh close to the falls of the Bow River." He also noted that "no other specimens were seen at this marsh, either in 1921 or 1926." Since then, I have found western red damsels at the Vermilion Lakes in Banff, a spring-fed stream just west of Bow Valley Provincial Park (it flows into the Bow River), a spring 19 km south of Stettler, right beside Highway 56, and at a badlands seep near the Jenner Bridge. Christine Rice and Jon Hornung discovered another population in 1999 at the Jenner Springs, also near Jenner.

I also looked into a report of a red-bodied damselfly near Bellevue, in the Crowsnest Pass, flying over running water passing over rocks, in about mid-June of 1997. The location turned out to be a spring stream pouring out of a culvert leading from the Bellevue Mine tour site, and it looks like good habitat for western red damsels. However, the original report also mentioned "solid black wings," which suggests no known damselfly in North America (none with red bodies at any rate). Let's keep an eye on that place!

At the Stettler spring, I have seen adults as early as May 29 (when they were old enough to mate) and as late as July 7. I also have a record of visiting the site on August 14 and finding none. At the Bow Valley location, I have seen them as late as July 25. A local couple told me that the Stettler spring has been flowing at least since the 1920s. Now, it comes out of a tap, and flows down a very short creek to a pond. Once, at the end of May, I measured the temperature at the tap at 3°C and the pond at 12°C. A few years later, on March 23, I found larvae in very shallow water, among thick aquatic vegetation, at the bottom of the slope where the creek trickles down. The temperature there was 6°C.

In June 2001, I found many pairs and many individuals while searching for tiger beetles around the small badlands seep near Jenner. There is almost no water at the site and most of it is in cow footprints, but I have known this spring since 1982. The water I tested had a pH of 8.6–9.3 and a conductivity of 3,100–6,100 micromhos/cm. In other words, it was pretty salty and alkaline, but no worse than some of the salt lakes mentioned above in the section on spreadwings.

A student of Gordon Pritchard's, Andrea Kortello, showed that although vivid dancers and western red damsels both inhabit springs, the vivid dancer is absent from open places since it needs trees as perching sites, while the western red damsel is absent from treed places for the opposite reason. Numerous authors, most notably Minter Westfall and Michael May, have referred to the hard-bottomed nature of the habitat of this species. I suppose this may be true in Alberta, but springs are a much more obvious requirement.

A bright orange-red female western red damsel, photographed at the Vermilion Lakes in Banff.

Which is the male? Two almost identical (but I promise, the one on the left is a female) sedge sprites in tandem.

Sedge Sprite

NEHALENNIA IRENE *("Nay-hah-lay-NAY-ah eye-REEN")*

There once was a damsel, Irene,
Who was glittering, glamorous green—
    Was a goddess her namesake,
    Or memories of heartbreak,
In Herr Dr. Hagen's late teens?

IDENTIFICATION

Males: These are very small, slender, metallic green damselflies with no postocular spots and a blue tip to the abdomen.
Females: Also very small and very green, with no postocular spots. Most females are male-like, with some blue markings, but some are green and yellow.
Length: About 27 mm.
Similar species: The eastern and western forktails are about the same size, but both of these species possess obvious postocular spots.

The Damselflies of Alberta 125

THE NAME

Nehalennia was the name of a goddess of Belgian Gaul, and Irene was the Greek and Roman goddess of peace. Hermann Hagen, who named this species in 1861, also described 61 other species in the North American odonate fauna, naming some of them Amanda, Elisa, Iris, and Zoe. It's hard to know if these all names refer to goddesses (and I doubt it, although anything is possible in the world of goddesses), since it was not customary in those days to explain new animal names. One prominent odonatist wrote to me (and wishes to remain anonymous), saying, "My guess is that there was a cute wench in Koenigsberg named Irene."

ECOLOGY

Habitat: Sedge sprites are found around ponds of all sorts: marshy ponds, beaver ponds, and marl ponds. They don't seem to do well at lakes, and adults are almost always found flying low to the ground amid grassy or sedge-rich vegetation.

Life history: On the wing from mid-June to mid-August, in most places.

Range: Probably found throughout Alberta, but not on the southeastern prairies.

Where to find them: Good habitat for this species is not hard to find, but the damsels themselves are tiny and difficult to spot. Try the Wagner Natural Area or the Opal Natural Area.

NOTES

Apparently, this is one of the most common damselflies in eastern Canada. Here in Alberta, however, it is not nearly as abundant nor as widespread. Edmund Walker wrote that "this species occurs locally in the prairie provinces," and perhaps that's the best way to think of it—as a localized species. Francis Whitehouse thought of it as a "prairie" species, oddly enough, even though we have precious few records of sedge sprites from the area we now know as the prairies.

There is no doubt that even in good habitat, this species is easy to miss where it occurs at low densities. The adults are so tiny and slender that they are actually hard to see (especially if you are 6 feet tall), and they fly inconspicuously among grasses and sedges most of the time. Still, once you get your search image tuned-in, you can survey for this damselfly without too much trouble, and in places where they appear to be rare, they probably really are rare. People from other parts of North America are more accustomed to tiny damselflies, and there are five species of sprites in North America, plus another in Europe.

Christine Rice did not find sedge sprites in her 1998 survey at Beaverhill Lake nor during odonate survey work she did between Medicine Hat and Brooks with Jon Hornung in 1999. Godo Stoyke also failed to find it at the Devonian Botanical Gardens, in 1986. In these three tidbits of negative data, we may have the key to this species' distribution in the province. Beaverhill Lake is probably just too big for sprites, since they prefer quiet ponds. The dry grasslands are also unable to produce good sprite habitat, for reasons that may involve dryness, too much wind, or salinity. But what about the Devonian Botanical Gardens, with their tranquil ponds and water-filled moats?

My guess is that the sedge sprite is a low-flying, rather sedentary species that may not have colonized the gardens quite yet. After all, the habitats there are all man-made. If this is true, the sedge sprite is a minor member of the "damselfly rain" that I discuss elsewhere in this book. The fact that Donald Hilton also failed to find sedge sprites in his survey of the Cypress Hills provides more evidence that this species does not wander far in search of new habitats, as there are many places in the hills where a sprite would likely feel at home.

On the other hand, shortly before this book went to press, I visited a spring-fed dugout on private ranchland north of Medicine Hat, with my students Lindsay Wickson and Randy Dzenkiw, and a visiting student from Mexico, Andreas Stange. Sure enough, we found a few sedge sprites in the grass, along with western red damsels, plains and western forktails, taiga bluets, four species of American bluets, and common spreadwings. This little pond, about the size of my house, may well prove to be the most diverse site for damselflies in the entire province once it is well-studied, and it is smack in the middle of the prairie zone where sedge sprites are most difficult to find.

A damselfly drawing by Matthew Brownoff.

APPENDIX 1

Checklist of Alberta Damselflies

This is a list of not only the species we have here in Alberta, but also all of the different-looking colour morphs, sexual dimorphisms, and colour changes affected by age.

DAMSELFLIES, suborder Zygoptera

   JEWELWINGS, family Calopterygidae
      river jewelwing, *Calopteryx aequabilis* Say

   SPREADWINGS, family Lestidae
      emerald spreadwing, *Lestes dryas* Kirby
      common spreadwing, *Lestes disjunctus* Selys
      lyre-tipped spreadwing, *Lestes unguiculatus* Hagen
      spotted spreadwing, *Lestes congener* Hagen

   POND DAMSELS, family Coenagrionidae
      prairie bluet, *Coenagrion angulatum* Walker
      taiga bluet, *Coenagrion resolutum* (Hagen)
      subarctic bluet, *Coenagrion interrogatum* (Hagen)
      boreal bluet, *Enallagma boreale* Selys
      northern bluet, *Enallagma cyathigerum* (Charpentier)
      Hagen's bluet, *Enallagma hageni* (Walsh)
      marsh bluet, *Enallagma ebrium* (Hagen)
      alkali bluet, *Enallagma clausum* Morse
      river bluet, *Enallagma anna* Williamson
      tule bluet, *Enallagma carunculatum* Morse
      Pacific forktail, *Ischnura cervula* Selys
      plains forktail, *Ischnura damula* Calvert
      western forktail, *Ischnura perparva* Selys
      eastern forktail, *Ischnura verticalis* (Say)
      vivid dancer, *Argia vivida* Hagen
      sedge sprite, *Nehalennia irene* (Hagen)
      western red damsel, *Amphiagrion abbreviatum* (Selys)

"Eyeballing" a damselfly is an enjoyable challenge, but if you really want to be sure of your identifications, try using the key that begins on the next page.

APPENDIX 2

Key to the Adult Damselflies of Alberta

Coupled with good illustrations, or comparative material, this key should work with any single specimen in the hand. In the field, habitat, range, and presence of both males and females together should make identification much easier and provide good double-checks on many of the trickier identifications.

Remember, a key is a way of organizing your thinking process. It contains no new information not found in simple descriptions of the species and no sure-fire "tricks" for identification. It still requires practice to use correctly, and the more familiar you are with the damselflies themselves, the more success you will have with the key.

1. Does it have postocular spots? (On some damselflies these spots are expanded so much that they join to the pale areas on the back of the head.) If so, go to 10.
Or are the postocular spots absent? If so, go to 2.

2. (from 1)
Does it have a pale shoulder stripe, and *no* orange or red colouration? If so, go to 3.
Or is it red or orange, or is the pale shoulder stripe absent? If so, go to 8.

3. (from 2)
Are there small, oval, dark blotches in the pale areas on the sides of the thorax, just above the bases of the hind legs? Then it is the **spotted spreadwing,** *Lestes congener*.
Or are these blotches absent (this area, the metepimeron, is entirely pale or mostly dark)? If so, go to 4.

4. (from 3)
Is it a male? If so, go to 5.
Or is it a female? If so, go to 6.

5. (from 4)
Are the lower claspers more or less straight when viewed from the top? Then it is the **common spreadwing**, Lestes disjunctus.
Or are the lower claspers strongly curved and lyre-shaped? Then it is the **lyre-tipped spreadwing**, Lestes unguiculatus.
Or are the lower claspers spoon-shaped? Then it is the **emerald spreadwing**, Lestes dryas.

6. (from 4)
Is the body relatively heavy and bright green in colour? Then it is the **emerald spreadwing**, Lestes dryas.
Or is the body slimmer and either a less-intense green or bronze in colour? If so, go to 7.

7. (from 6)
Is the back of the head mostly dark, especially to the sides, behind the main part of the eyes? Then it is the **common spreadwing**, Lestes disjunctus.
Or are there large pale areas on the back of the head? Then it is the **lyre-tipped spreadwing**, Lestes unguiculatus.

8. (from 2)
Are the wings dark near the tip and more clear near the base? If so, it is the **river jewelwing**, Calopteryx aequabilis.
Or are the wings free of dark areas? If so, go to 9.

9. (from 8)
Is the body red or orange, with black markings? Then it is the **western red damsel**, Amphiagrion abbreviatum.
Or is the body at least partly iridescent green? Then it is the **sedge sprite**, Nehalennia irene.

10. (from 1)
Is it a male? In other words, does it have claspers at the tip of its abdomen? If so, go to 11.
Is it a female, with an ovipostor blade on the underside of the tip of the abdomen? If so, go to 26.

11. (from 10)
Are there small, dark markings on the sides of the abdomen, near the rear of each long segment, separate from those on the top? And are the spines on the tibiae (the outermost long segments of the legs) longer than the spaces between their bases? Then it is the **vivid dancer**, Argia vivida.
Or are the dark markings not like this, and the spines shorter? If so, go to 12.

12. (from 11)
    Is the top of the pterothorax marked with four pale spots on a dark background? If so, go to 13.
    Or is the top of the thorax marked with pale shoulder stripes or exclamation marks on a dark background? If so, go to 14.

13. (from 12)
    Is the bump on the top of abdominal segment 10 relatively high and distinctly notched? Then it is the **Pacific forktail,** *Ischnura cervula.*
    Or is the bump on the top of the tenth abdominal segment relatively low and not noticeably notched? Then it is the **plains forktail,** *Ischnura damula.*

14. (from 13)
    Are the pale areas of the body and the eyes both blue and green, or blue and blue-green? If so, go to 15.
    Or are the pale areas of the body and the eyes blue, without obvious areas that are green or blue-green? If so, go to 18.

15. (from 14)
    Is the abdomen blue-tipped and otherwise quite dark in colour? If so, go to 16 (and note that these damselflies are usually very small in size).
    Or is the abdomen either dark-tipped or with obvious blue areas near the base? If so, go to 17.

16. (from 15)
    Are the spots on the top of the head small, and is the prothorax black when viewed from above? Then it is the **western forktail,** *Ischnura perparva.*
    Or are the spots on the top of the head large, and are there two small pale spots on the prothorax when viewed from above? Then it is the **eastern forktail,** *Ischnura verticalis.*

17. (from 15)
    Are segments 3–5 of the abdomen almost entirely dark when viewed from above? Then it is the **prairie bluet,** *Coenagrion angulatum.*
    Or do segments 3–5 of the abdomen have blue bases? Then it is the **taiga bluet,** *Coenagrion resolutum.*

18. (from 14)
    Are the first three dark bands on the abdomen about equal in length? If so, go to 19.

Or do the first three dark bands on the abdomen increase in length from the base toward the tip? If so, go to 21.

19. (from 18)
Are the first three dark bands on the abdomen relatively short (half as long or less than the longest dark bands on the abdomen)? If so, go to 20.
Or are the first three dark bands on the abdomen longer, and are the postocular spots of medium size? Then it is the **alkali bluet**, *Enallagma clausum* (but be sure to check the shape of the claspers as well).

20. (from 19)
Do the upper claspers, when viewed from the side, look like mittens, with a thumblike projection? Then it is the **northern bluet**, *Enallagma cyathigerum*.
Or is the thumblike projection absent, or very small and not at all obvious? Then it is the **boreal bluet**, *Enallagma boreale*.

21. (from 18)
Are the upper claspers C-shaped in side view? Then it is the **marsh bluet**, *Enallagma ebrium*.
Or are the upper claspers not C-shaped? Then go to 22.

22. (from 21)
Are the upper claspers long, bluntly pointed, and extended somewhat upward? Was it found near a stream? Then it is the **river bluet**, *Enallagma anna*.
Or are the upper claspers shorter and not as described above? Then go to 23.

23. (from 22)
Are the upper claspers blunt in side view and longer than the lower claspers? Then it is the **tule bluet**, *Enallagma carunculatum*.
Or are the upper claspers shorter than the lower claspers? If so, go to 24.

24. (from 23)
Are the upper claspers pointed and triangular in side view? Then it is the **Hagen's bluet**, *Enallagma hageni*.
Or are the upper claspers more rounded? If so, go to 25.

25. (from 24)
Do the upper claspers, when viewed from the side, look like mittens, with a thumblike projection? Then it is the **northern bluet**, *Enallagma cyathigerum*.

Or is the thumblike projection absent, or very small and not at all obvious? Then it is the **boreal bluet,** *Enallagma boreale*

26. (from 10)
Are the spines on the tibiae (the outermost long segment of the legs) longer than the spaces between their bases, and are there dark markings among the sides of the long abdominal segments, as well as along the tops? Then it is the **vivid dancer,** *Argia vivida*.
Or are the spines on the tibiae not as long as the spaces between their bases, and are the dark abdominal markings confined to the top of the long segments? If so, go to 27.

27. (from 26)
Does the median vein branch somewhere below the fourth crossvein past the nodus (the notch along the leading edge of the wing) on the forewing, and the second or third on the hind wing? If it is an older female, are its colours obscured by darkening or bluish-white powder? If so, go to 28.
Or does the median vein branch somewhere below the fifth crossvein past the nodus on the forewing, and the fourth (or beyond) on the hind wing? If it is an older female, are its colours clear and unobscured? If so, go to 31.

28. (from 27)
Is the underside of the eighth abdominal segment smooth? Then it is the **western forktail,** *Ischnura perparva*.
Or does the underside of the eighth abdominal segment protrude as a spine? If so, go to 29.

29. (from 28)
Is the damselfly small (about 25 mm), are the pale shoulder stripes unbroken, and are the hind margins of the shoulder pads low and not raised into lobes? Then it is the **eastern forktail,** *Ischnura verticalis*.
Or is the damselfly larger (about 30 mm) with the hind margins of the shoulder pads raised into lobes (and in some individuals, the pale shoulder stripes are broken to form two dots each)? If so, go to 30.

30. (from 29)
Does it lack nipple-like bumps on the top of the prothorax and possess a blunt knob on the rear of the prothorax, facing backward? Then it is the **Pacific forktail,** *Ischnura cervula*.
Does it have nipple-like bumps on the top of the prothorax and lack a blunt knob on the rear of the prothorax, facing backward? Then it is the **plains forktail,** *Ischnura damula*.

31. (from 27)
	Is the underside of the eighth abdominal segment smooth? If so, go to 32.
	Or does the underside of the eighth abdominal segment protrude as a spine? If so, go to 34.

32. (from 31)
	Is the pale shoulder stripe broken into elongate and dotlike portions, to form an exclamation mark? Then it is the **subarctic bluet**, *Coenagrion interrogatum*.
	Or is the pale shoulder stripe entire? If so, go to 33.

33. (from 32)
	Is the abdomen almost entirely dark when viewed from above, apart from very narrow light bands on some segments? Then it is the **taiga bluet**, *Coenagrion resolutum*.
	Or is the abdomen banded with at least one obvious light (usually blue) patch? Then it is the **prairie bluet**, *Coenagrion angulatum*.

34. (from 31)
	Is the abdomen almost entirely dark when viewed from above, with very thin, light rings at the junctions of the abdominal segments? If so, go to 35.
	Or are there wider, pale rings between the dark areas, and in some individuals a pale patch near the tip of the abdomen as well? If so, go to 36.

35. (from 34)
	Do the shoulder pads have a raised hind margin? Then it is the **marsh bluet**, *Enallagma ebrium*.
	Or do the shoulder pads lack a raised hind margin? Then it is the **Hagen's bluet**, *Enallagma hageni*.

36. (from 34)
	Is there a pair of pits on the pronotum? Then it is the **alkali bluet**, *Enallagma clausum*.
	Or are there no pits on the pronotum? If so, go to 37.

37. (from 36)
	Are the postocular spots large? If so, go to 38.
	Or are the postocular spots of moderate size? If so, go to 39.

38. (from 37)
	Are the sides of the triangular dark patch on abdominal segment 8 straight (if a dark patch is present at all), and is the groove at the base of the

shoulder pads continuous across their width? Then it is the **northern bluet,** *Enallagma cyathigerum*.

Or are the sides of the triangular dark patch on abdominal segment 8 concave (if a dark patch is present at all), and does the groove at the base of the shoulder pads extend only part way across their width? Then it is the **boreal bluet,** *Enallagma boreale*.

39. (from 37)

    Do the shoulder pads match the **river bluet** on page 156? (It may also be the case that the dark shoulder stripe is very thin, the top of segment 8 is at least half pale, and the damselfly was found near a prairie stream.) Then it is the **river bluet,** *Enallagma anna*.

    Or do the shoulder pads look more like those of the **tule bluet** (and the dark shoulder stripes are usually of average width, and the top of segment 8 is mostly dark)? Then it is the **tule bluet,** *Enallagma carunculatum*.

Most damselfly netting is done near the ground, but the occasional high-flier requires an extra effort to catch!

APPENDIX 3

Helpful Sources for Damselfly Study

SUPPLIES

Bio Quip Inc.: a good selection of nets, other entomological equipment, books and videos.
address: 17803 LaSalle Avenue, Gardena, California 90248-3602
phone: (310) 324 0620   *fax*: (310) 324 7931   *email*: bioquip@aol.com

Jean Paquet: insect nets and collection supplies.
address: 3 reu du Coteau, P.O. Box 953, Pont Rouge, Québec GH3 2E1
email: jeanpaquet@webnet.qc.ca

JOURNALS, WEBSITES, AND SOCIETIES

Dragonfly Society of the Americas: a group of amateur and professional odonatologists that meets as a whole (and as regional sections) on a regular basis and publishes the newsletter *Argia* and the more formal *Bulletin of American Odonatology*.
address: c/o T. Donnelly, 2091 Partridge Lane, Binghamton, New York 13903
website: http://www.afn.org/~iori/dsaintro.html

Entomological Society of Alberta: a group of entomologists (primarily professional) that meets once a year for three days and publishes abstracts from papers presented at the meetings. Amateurs are welcome. Membership is $10 per year ($5 for students) payable to the Secretary of the Entomological Society of Alberta.
address: c/o Department of Biological Sciences, University of Alberta, Edmonton, Alberta T6G 2E9

Entomological Society of Canada: a group of entomologists that meets once a year and publishes a newsletter as well as *The Canadian Entomologist*. Amateurs are welcome. Membership fee depends on whether you live in Canada or not, are a student or not, and which publications you wish to receive.
address: 1320 Carling Avenue , Ottawa, Ontario K1Z 7K9
website: http://www.biology.ualberta.ca/esc.hp/homepage.htm

The E. H. Strickland Entomological Museum: houses a research collection of damselflies and other insects, and has a great website.
address: Room 218, Earth and Atmospheric Sciences Building, University of Alberta, Edmonton, Alberta   T6G 2E1
website: http://www.biology.ualberta.ca/uasm.html

Glossary

abdomen: The hindmost of the three main body parts of an insect, usually slender and elongate in damselflies, made up of ten segments.
antenna: A segmented feeler that extends from the forehead of an insect, separate from the mouthparts.
antennae: Plural of *antenna*.
aquatic: Living in water.
biodiversity: The diversity of life on earth, including species numbers as well as general measures of genetic diversity and ecosystem diversity.
biologist: Someone who studies living things.
cerci: See *upper claspers*.
class: In classification, a grouping of orders.
cold springs: Spring-fed waters that are cold to the human sense of touch.
cryptic: Coloured in a fashion that enhances camouflage.
dark shoulder stripe: The stripe that runs diagonally across the side of the *pterothorax* of many but not all damselflies.
database: A computer-based collection of information on the occurrence of individual damselflies in time and space.
diapause: A period of dormancy, which is necessary in order for an insect to complete its development to adulthood.
dispersal: A movement away from the place where a damselfly first emerges as an adult.
DNA analysis: Comparison of the sequence of the component parts of equivalent fragments of the genetic material (DNA—deoxyribose-nucleic acid) of damselflies, in order to estimate how closely different species are related.
ecology: The study of how organisms interact with their surroundings, both living and nonliving.
electrophoresis: A method for distinguishing different forms of the same enzyme (*enzymes* are compounds that catalyze chemical reactions within the body) which are presumed to correspond with different forms (*alleles*) of the same gene.
endocrine: The system of glands that produce *hormones* (chemicals important in growth, behaviour, and reproduction).
entomology: The study of insects.

evolutionary tree: A diagram showing how various evolutionary lineages are related. The tree shape comes from the fact that all lineages can trace their ancestry back to a common "trunk" in any given group of related organisms.

family: In classification, a grouping of genera.

faunistic: Of or relating to studies (like this book) that focus on the occurrence and ecology of the species in a particular group of related organisms that live in a particular area.

fecund: Capable of producing a great many young.

fluctuating asymmetry: The notion that male animals that are mirror images of themselves on the right and left side of the centre line are likely to be genetically superior to those that are not, and that females can use this to judge the quality of potential mates.

genera: Plural of *genus*.

genus: In classification, a grouping of species.

geographic race: A subdivision of a species, in which the animals in a particular geographic area share recognizable differences from those elsewhere; the same as *subspecies*.

gill: An organ that allows an animal to obtain oxygen from water and release waste gasses into water.

habitat: The sort of place in which an organism lives.

larva: An immature insect, not yet an adult.

Latinized words: Words constructed in accordance with the rules of Latin grammar, even if they are derived from non-Latin roots.

lower claspers: The lowermost pair of short appendages at the tip of a male damselfly's abdomen, used to grasp the female's prothorax during mating; the same as *inferior abdominal appendages*, or *paraprocts*.

marsh: A shallow, usually open and sunny wetland, largely vegetated by emergent plants with their roots under water and their stems and leaves above the surface.

mesostigmal laminae: See *shoulder pad*.

micromhos/cm: MicroSiemens per centimetre; a measure of the electrical conductivity of water, which increases with the amount of dissolved salts and is close to zero in distilled water.

molecular diagnostics: A reference to electrophoresis and/or DNA analysis.

nocturnal: Active by night.

NSERC: The National Science and Engineering Research Council, which is a primary source of funding for biology research in Canada.

ode: Short for *odonate*.

odester: Slang for *odonatologist*.

Odonata: The insect order that includes damselflies and dragonflies.

odonate: Colloquial name for members of the Odonata.

odonatist: The same as *odonatologist*.

odonatologist: A scientist who studies damselflies and dragonflies.

order: In classification, a grouping of families.

organism: A living thing.

ovipositor: The egg-laying organ of many female animals, including damselflies.

pale shoulder stripe: The pale stripe that runs diagonally across the side of the pterothorax of many but not all damselflies.

Paleoptera: Insects possessing wing muscles that directly attach to the wing bases (the "paleopterous" condition). Among living insects this includes only the dragonflies, damselflies, and mayflies.

paraprocts: See *lower claspers*.

parasite: An organism that feeds on the tissues of another organism without necessarily killing the "host."

peatland: A wetland in which the decomposition of plants is slower than the accumulation of dead plants, such that peat is formed on the ground surface. The usual sorts of peatlands in Alberta are called bogs and fens.

permanent ponds: Ponds that have water (or ice) in them all year.

petiole: The narrow stalk at the base of the wings of all of our Alberta damselflies except for the river jewelwing.

physical gill: A bubble or layer of air in contact the tracheal system of an aquatic insect, which allows the exchange of oxygen and waste gases between the insect and the surrounding water.

pond: A relatively small body of nonflowing water with a relatively large area of open water in its centre.

postocular spots: The pale spots behind the compound eyes and on top of the head of adult damselflies.

power (as in "10 power"): The amount of magnification provided by a lens, measured as the difference between the real and apparent length of an object; e.g., if the object appears ten times longer than it really is, then the lens is "10 power" or "10X."

prairie: The grassland region, which in Alberta extends south from about the latitude of Hannah and east from about the longitude of Calgary.

predatory: Habitually feeding on animals, which are killed and then eaten.

prereproductive period: The time in a damselfly's life between emerging as an adult and first mating.

pronotum: The upper surface of the prothorax.

prothorax: The first of the three segments that make up the thorax (the one closest to the head). In damselflies it is small and looks like a neck.

pruinosity: On some adult odonates, a light greyish or blue pigment that is deposited on the outside of the body once the damselfly reaches adulthood; also called *pruinescence*, and a damselfly with such pigment is said to be *pruinose*.

pterothorax: The fused second and third segments of the thorax, that bear the wings. In this book, I use the word *thorax* to refer to the *pterothorax*, while referring to the *prothorax* or *pronotum* by name.

reproductive period: The period in an adult damselfly's life during which it is able to mate and/or lay eggs.

sabbatical: For a university professor, a year-long period free from the need to teach or to remain on campus.

secondary adaptation: An attribute of an organism that evolved once, was lost, and then evolved again, albeit usually in a somewhat different form.

secondary sex organs: Any organs of reproduction other than the testes, ovaries, and mating organs at the tip of the abdomen. In damselflies, this term generally refers to the temporary sperm storage organ at the base of a male's abdomen.

segment: A component of a body part or of the body itself. Usually segments are lined up one behind the other and are separated from one another by a narrow ring of soft tissue, rather than the usual hardened exterior cuticle of an insect. On a damselfly, the most obvious examples are antennal segments, leg segments, and abdominal segments.

shoulder pad: The low, padlike ridges that run across the top front of the pterothorax of female damselflies.

slough: As far as I'm concerned, the same as a marsh, although most people give the term an overtone of disgust.

species: A group of organisms that not only look more or less alike, but are capable of interbreeding in nature and which form a more or less cohesive evolutionary lineage.

spermatheca: A structure in the reproductive system of many female insects that allows them to store sperm from a previous mating and use it later to fertilize their eggs.

spring-fed ponds: Ponds that are kept filled by groundwater. In Alberta, these sorts of ponds rarely become completely ice-covered in winter.

stalked wings: Wings with *petioles*.

stigmata: The distinctive cells near the outer leading edges of each of an odonate's wings, which are usually darker or lighter than the rest of the wing and apparently help with the aerodynamics of flight.

structural: Having to do with the anatomy of an organism.

subclass: In classification, a group of orders distinguished from other such subclasses within a given class. Not all classes are divided into subclasses.

suborder: In classification, a group of families distinguished from other such suborders within a given order. Not all orders are divided into suborders.

subspecies: The same as *geographic race*.

systematics: The scientific study of how animals are related in an evolutionary sense and how this should be reflected in the way they are classified.

taxonomy: An older term for *systematics*, without the same emphasis on evolutionary relationships.

temporary ponds: Ponds that dry up during part of the year, usually from mid- to late summer until the following spring.

thorax: The second main body part of an insect, between the head and the abdomen, and comprising the prothorax and the pterothorax.

upper claspers: The uppermost pair of claspers on a male damselfly's abdomen, used to grasp the female's prothorax during mating; the same as *superior abdominal appendages*, or *cerci*.

venation: The pattern of "veins" in an insect wing. These veins are important in the process of allowing the wing to harden when the insect emerges, but are not like our own veins in that they do not carry blood back to the heart.

warm springs: Spring-fed waters that are warm to the human sense of touch.

References

Acorn, John H. 1983. New distribution records of Odonata from Alberta, Canada. *Notulae Odonatologicae.* 2(2): 17–32.

———. 1995. "Acorn, the Nature Nut." Episode no. 20: *Dragonflies*. Great North Productions, Edmonton. Video: 22 minutes.

———. 2001. *Tiger Beetles of Alberta: Killers on the Clay, Stalkers on the Sand.* Alberta Insects Series. University of Alberta Press, Edmonton. 120 pp.

Baker, Robert, and Hugh Clifford. 1980. The nymphs of *Coenagrion interrogatum* and *C. resolutum* (Zygoptera: Coenagrionidae) from the boreal forest of Alberta, Canada. *The Canadian Entomologist.* 112: 433–436.

Bick, G. H., and L. E. Hornuff. 1965. Behaviour of the damselfly *Lestes unguiculatus* Hagen (Odonata: Lestidae). *Proceedings of the Indiana Academy of Sciences.* 75: 110–115.

———. 1966. Reproductive behaviour in the damselflies *Enallagma aspersum* (Hagen) and *Ennalagma exsulsans* (Hagen) (Odonata: Coenagrionidae). *Proceedings of the Entomological Society of Washington.* 68: 78–85.

Brown, A. F., and D. Pascoe. 1988. Studies on the acute toxicity of pollutants to freshwater macroinvertebrates. 5. The acute toxicity of cadmium to twelve species of predatory macroinvertebrates. *Archives of Hydrobiology.* 114: 311–319.

Brown, Jonathan M., Mark A. McPeek, and Michael L. May. 2000. A phylogenetic perspective on habitat shifts and diversity in the North American *Enallagma* damselflies. *Systematic Biology.* 49: 697–712.

Cannings, Robert A. 1982. Obituary: George P. Doerksen. *Odonatologica.* 11: 59–62.

———. 1983. Odonata collected in Banff National Park, Alberta, Canada, during the Seventh International Symposium of Odonatology. *Notulae Odonatologicae.* 2: 23–24.

———. 1984. Noteworthy Odonata collected by participants in the Seventh International Symposium of Odonatology. *Notulae Odonatologicae.* 2: 53–55.

Cannings, Robert A., and George P. Doerksen. 1979. Description of the larva of *Ischnura erratica* (Odonata: Coenagrionidae) with notes on the species in British Columbia. *The Canadian Entomologist.* 111: 327–331.

Cannings, Robert A., and Kathleen Stuart. 1977. *The Dragonflies of British Columbia*. Handbooks of the British Columbia Provincial Museum. 35. 254 pp.

Cannings, Robert A., Sydney G. Cannings, and Richard J. Cannings. 1980. The distribution of the genus *Lestes* in a saline lake series in British Columbia, Canada. *Odonatologica*. 9: 19–28.

Cannings, Sydney G., and Robert A. Cannings. 1980. The larva of *Coenagrion interrogatum* (Odonata: Coenagrionidae) with notes on the species in the Yukon. *The Canadian Entomologist*. 112: 437–441.

———. 1994. The Odonata of the northern cordilleran peatlands of North America. *Memoirs of the Entomological Society of Canada*. 169: 89–110.

———. 1997. Dragonflies (Odonata) of the Yukon. In H. V. Danks and J. A. Downs, eds. *Insects of the Yukon*. Biological Survey of Canada (Terrestrial Arthropods). Ottawa. pp 170–200.

Cannings, Sydney G., Robert A. Cannings, and Richard J. Cannings. 1991. Distribution of the dragonflies (Insecta: Odonata) of the Yukon Territory, Canada, with notes on ecology and behaviour. No. 13, *Contributions to Natural Science*, Royal British Columbia Museum.

Conrad, Kelvin F. 1992. Relationships of larval phenology and imaginal size to male pairing success in *Argia vivida* Hagen (Zygoptera: Coenagrionidae). *Odonatologica*. 21: 335–342.

Conrad, Kelvin F., and Gordon Pritchard. 1988. The reproductive behaviour of *Argia vivida* Hagen: an example of a female control mating system (Zygoptera: Coenagrionidae). *Odonatologica*. 17: 179–185.

———. 1989. Female dimorphism and physiological colour change in the damselfly *Argia vivida* Hagen (Odonata: Coenagrionidae). *Canadian Journal of Zoology*. 67: 298–304.

Corbet, Philip S. 1999. *Dragonflies: Behavior and Ecology of Odonata*. Cornell University Press, Ithaca. 829 pp.

Cruden, Robert W., and O. J. Gode, Jr. 2000. The Odonata of Iowa. *Bulletin of American Odonatology*. 6(2): 13–48.

Donnelly, Nick. 2000. Dot-map project—hung up on *Lestes*! *Argia*. 12(3): 31.

Finke, Ola M. 1982. Lifetime mating success in a natural population of the damselfly *Enallagma hageni* (Walsh) (Odonata: Coenagrionidae). *Behavioral Ecology and Sociobiology*. 10: 293–302.

———. 1986. Underwater oviposition in a damselfly (Odonata: Coenagrionidae) favors male vigilance, and multiple mating by females. *Behavioral Ecology and Sociobiology*. 18: 405–412.

Forbes, M., B. Leung, and G. Schalk. 1997. Fluctuating asymmetry in *Coenagrion resolutum* (Hagen) in relation to age and male pairing success (Zygoptera: Coenagrionidae). *Odonatologica*. 26: 9–16.

Forbes, M. R. L., and Robert L. Baker. 1991. Condition and fecundity of the damselfly *Enallagma ebrium* (Hagen): the importance of ectoparasites. *Oecologia.* 86: 335–341.

Griffiths, Graham C. D., and Dierdre Griffiths. 1980. Preliminary insect survey of the Clifford E. Lee Nature Sanctuary (Alberta) during 1980. Unpublished report.

Hilton, Donald F. J. 1981. Flight periods of Odonata inhabiting a black spruce–sphagnum bog in south-eastern Quebec, Canada. *Notulae Odonatologicae.* 1: 127–130.

———. 1985. Dragonflies (Odonata) of Cypress Hills Provincial Park, Alberta and their biogeographic significance. *The Canadian Entomologist.* 117: 1127–1136.

Holder, Matt. 1998. *Lestes forcipatus* oviposition. *Argia.* 9 (4):20.

Johnson, Clifford. 1964. The inheritance of female dimorphism in the damselfly, *Ischnura damula. Genetics.* 49: 513–519.

Johnson, D. M., and P. H. Crowley. 1980. Habitat and seasonal segregation among coexisting odonate larvae. *Odonatologica.* 9: 297–308.

Korbut, Fred. 1997. River Jewelwing damselfly. *Alberta Naturalist.* 27 (1): 7.

Leggot, Mark, and Gordon Pritchard. 1985a. The effect of temperature on rate of egg and larval development in populations of *Argia vivida* Hagen (Odonata: Coenagrionidae) from habitats with different thermal regimes. *Canadian Journal of Zoology.* 63: 2578–2582.

———. 1985b. The life cycle of *Argia vivida* Hagen: developmental types, growth ratios and instar identification (Zygoptera: Coenagrionidae). *Odonatologica.* 14: 201–210.

———. 1986. Thermal preference and activity thresholds in populations of *Argia vivida* (Odonata: Coenagrionidae) from habitats with different thermal regimes. *Hydrobiologia.* 140: 85–92.

Martens, A. 1994. Field experiments on aggregation behaviour and oviposition in *Coenagrion puella* (L.) (Zygoptera: Coenagrionidae). *Advances in Odonatology.* 6: 49–58.

McPeek, Mark A. 1998. The consequences of changing the top predator in a food web: a comparative experimental approach. *Ecological Monographs.* 68(1): 1–23.

———. 2000. Predisposed to adapt? Clade-level differences in characters affecting swimming performance in damselflies. *Evolution.* 54: 2072–2080.

McPeek, Mark A., and Jonathan M. Brown. 2000. Building a regional species pool: diversification of the *Enallagma* damselflies in eastern North America. *Ecology.* 81: 904–920.

Miller, P. L., and C. A. Miller. 1981. Field observations on copulatory behaviour in Zygoptera, with an examination of the structure and activity of the male genitalia. *Odonatologica.* 10: 201–218.

More, Gavin. 1999. Damselflies and Dragonflies: an overview of the status and distribution of odonates in Alberta. Prepared by 49 North Creative Learning and Training for Parks Canada.

Page, Natasha. 1998. 1995 odonate survey at the Wagner Natural Area. *Alberta Naturalist*. 28: 61–64.

Paulson, Dennis R. 1974. Reproductive isolation in damselflies. *Systematic Zoology*. 23: 40–49.

Perry, S. J., and P. L. Miller. 1991. The duration of the stages of copulation in *Enallagma cyathigerum* (Charpentier) (Zygoptera: Coenagrionidae). *Odonatologica*. 20: 349–355.

Pilon, J.-G. 1981. Influence of temperature on the embryonic development of *Enallagma vernale* (Gloyd) and *E. ebrium* (Hagen) (Odonata: Coenagrionidae) in Quebec. *Abstracts of the 6th International Symposium of Odonatology*. 38.

Pritchard, Gordon. 1971. *Argia vivida* Hagen (Odonata: Coenagrionidae) in hot pools at Banff. *Canadian Field Naturalist*. 85: 186–188.

———. 1988. Dragonflies of the Cave and Basin Hot Springs, Banff National Park, Alberta, Canada. *Notulae Odonatologicae*. 3: 1–16.

Pritchard, Gordon, and Adrea Kortello. 1997. Roosting, perching, and habitat selection in *Argia vivida* Hagen and *Amphiagrion abbreviatum* (Selys) (Odonata: Coenagrionidae), two damselflies inhabiting geothermal springs. *Canadian Entomologist*. 129: 733–743.

Pritchard, Gordon, and Brian Pelchat. 1977. Larval growth and development of *Argia vivida* (Odonata: Coenagrionidae) in warm sulphur pools at Banff, Alberta. *Canadian Entomologist*. 109: 1563–1570.

Rehfeldt, Gunnar E. 1991. The upright male position during ovipostition as an anti-predator response in *Coenagrion puella* (L.) (Zygoptera: Coenagrionidae). *Odonatologica*. 20: 69–74.

Rice, Christine. 1999. Odonates at Beaverhill Lake. *Alberta Naturalist*. 29 (2): 37–44.

———. 2000. Status of Alberta Wildlife 2000: Preliminary status evaluation of the odonates: dragonflies and damselflies. Prepared for Fisheries and Wildlife Management Division, Natural Resource Service, Alberta Environment.

Robert, Adrien. 1963. *Les Libellules du Québec*. Service de la Faune, Bulletin 1.

Sawchyn, W. W., and C. Gillott. 1974a. The life history of *Lestes congener* (Odonata: Zygoptera) on the Canadian prairies. *Canadian Entomologist*. 106: 367–376.

———. 1974b. The life histories of three species of *Lestes* (Odonata: Zygoptera) in Saskatchewan. *Canadian Entomologist*. 106: 1283–1293.

———. 1975. The biology of two related species of coenagrionid dragonflies (Odonata: Zygoptera) in western Canada. *Canadian Entomologist*. 107: 119–128.

Sebastien, A., M. M. Sein, M. M. Thu, and P. S. Corbet. 1990. Suppression of *Aedes aegypti* (Diptera: Culicidae) using augmentative release of dragonfly larvae (Odonata: Libellulidae) with community participation in Yangon, Myanmar. *Bulletin of Entomological Research.* 80: 223–232.

Stoyke, Godo. 1987. Dragonflies of the Devonian Botanic Garden. *Alberta Naturalist.* 17 (2): 49–54.

Waage, Jonathan K. 1979. Dual function of the damselfly penis: sperm removal and transfer. *Science.* 203: 916–918.

Walker, Edmund M. 1953. *The Odonata of Canada and Alaska. Volume 1.* University of Toronto Press. Toronto, Ontario.

———. 1958. *The Odonata of Canada and Alaska. Volume 2.* University of Toronto Press. Toronto, Ontario.

Walker, Edmund M., and Philip S. Corbet. 1978. *The Odonata of Canada and Alaska. Volume 3.* University of Toronto Press. Toronto, Ontario.

Westfall, Minter J., and Michael L. May. 1996. *Damselflies of North America.* Scientific Publishers.

Whitehouse, Francis C. 1918. *Dragonflies (Odonata) of Alberta.* Alberta Natural History Society, Red Deer.

———. 1941. British Columbia dragonflies (Odonata) with notes on distribution and habits. *American Midland Naturalist.* 26: 488–557.

Wiggins, Glenn B., ed. 1966. *Centennial of Entomology in Canada, 1863–1963: A tribute to Edmund M. Walker.* Contribution No. 69, Life Sciences, Royal Ontario Museum, University of Toronto. University of Toronto Press. 94 pp.

A Gallery of Damselflies

adult male — adult female

RIVER JEWELWING
Calopteryx aequabilis, p. 54

young adult female — young adult male — mature adult male — young adult female — young adult male — mature adult male

EMERALD SPREADWING
Lestes dryas, p. 60

COMMON SPREADWING
Lestes disjunctus, p. 63

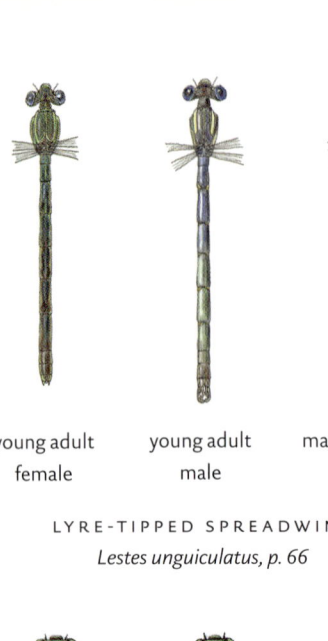

| young adult female | young adult male | mature adult male | | young adult female | young adult male | mature adult male |

LYRE-TIPPED SPREADWING
Lestes unguiculatus, p. 66

SPOTTED SPREADWING
Lestes congener, p. 69

| male | blue female | green female | | male | blue female | green female |

PRAIRIE BLUET
Coenagrion angulatum, p. 75

TAIGA BLUET
Coenagrion resolutum, p. 78

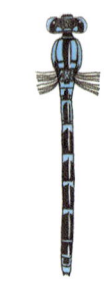

| male | blue female | green female | | typical male | dark male | yellow-green female |

SUBARCTIC BLUET
Coenagrion interrogatum, p. 81

BOREAL BLUET
Enallagma boreale, p. 86

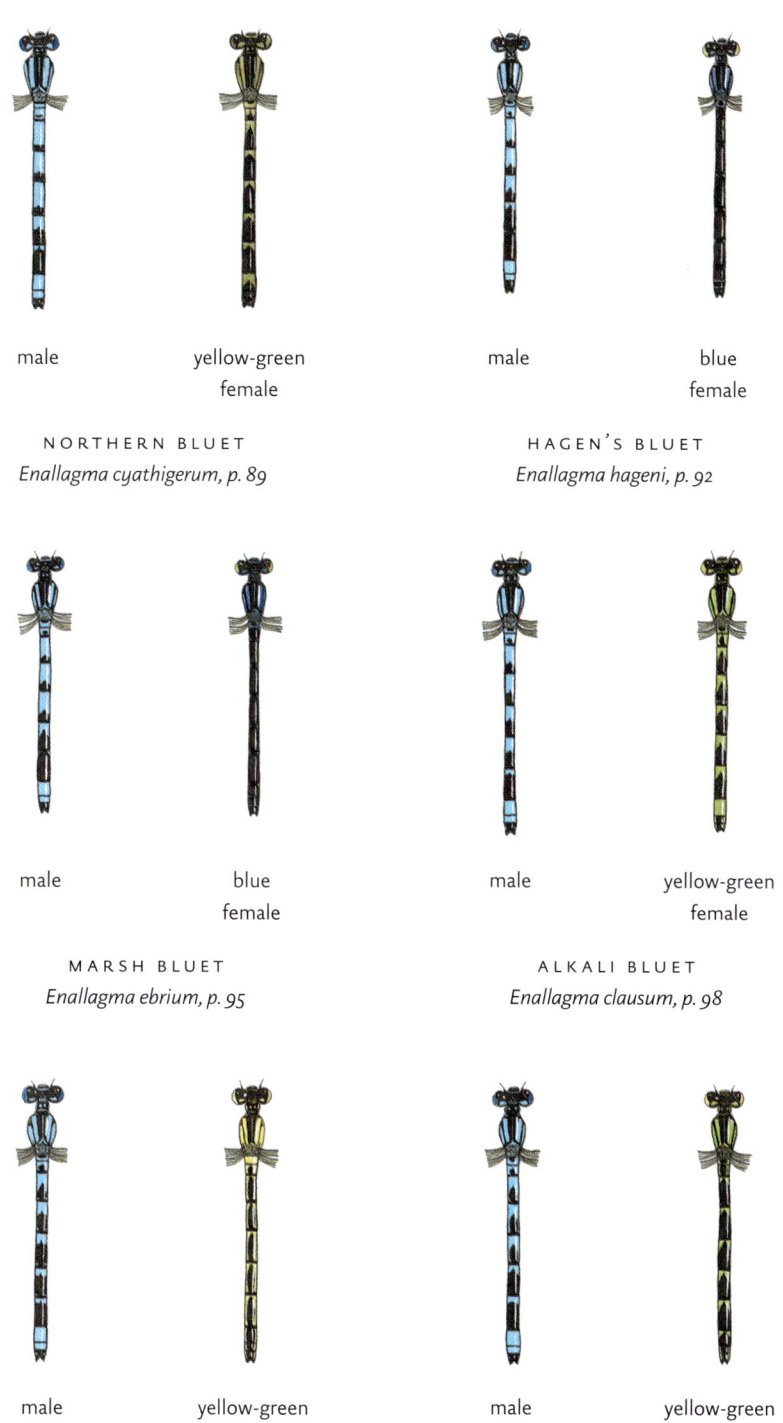

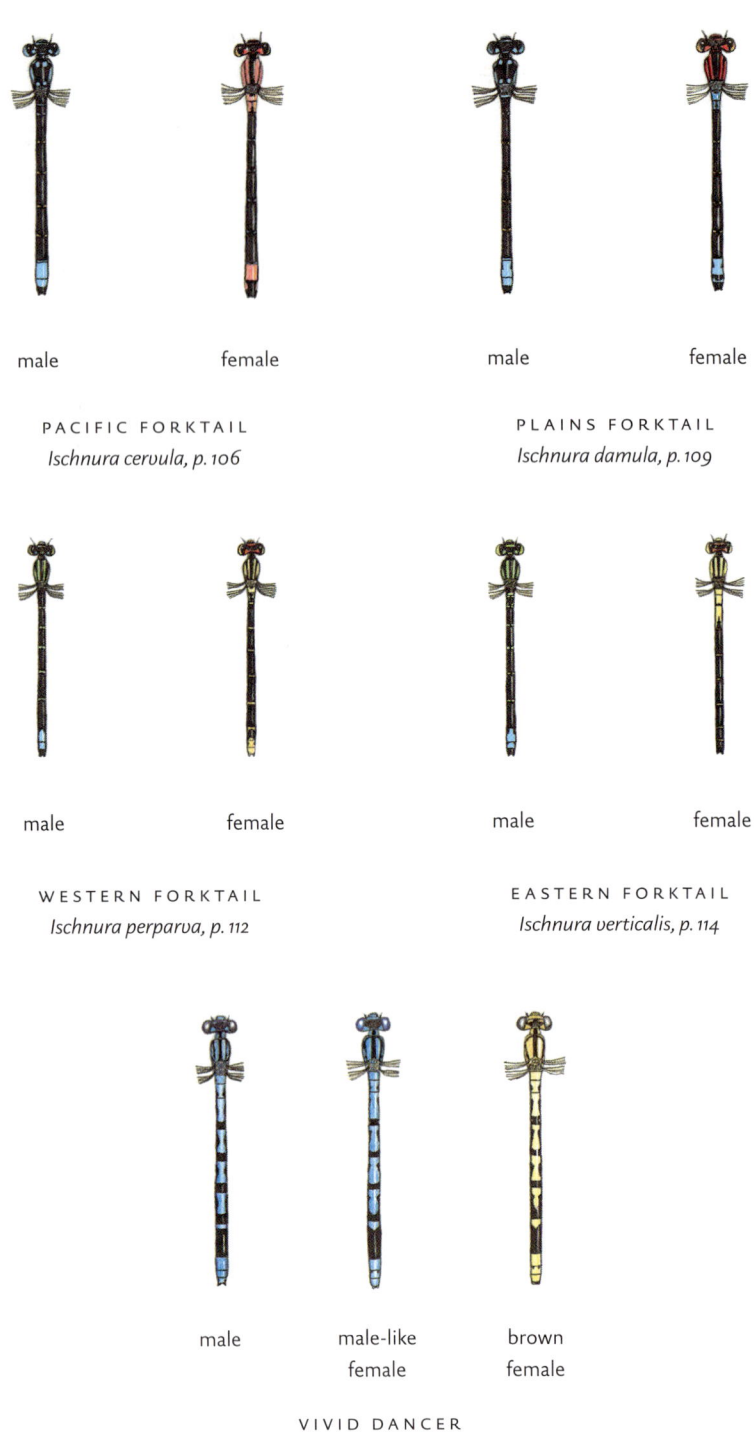

male　　　　female　　　　　　　male　　　　female

PACIFIC FORKTAIL
Ischnura cervula, p. 106

PLAINS FORKTAIL
Ischnura damula, p. 109

male　　　　female　　　　　　　male　　　　female

WESTERN FORKTAIL
Ischnura perparva, p. 112

EASTERN FORKTAIL
Ischnura verticalis, p. 114

male　　male-like　　brown
　　　　female　　　female

VIVID DANCER
Argia vivida, p. 118

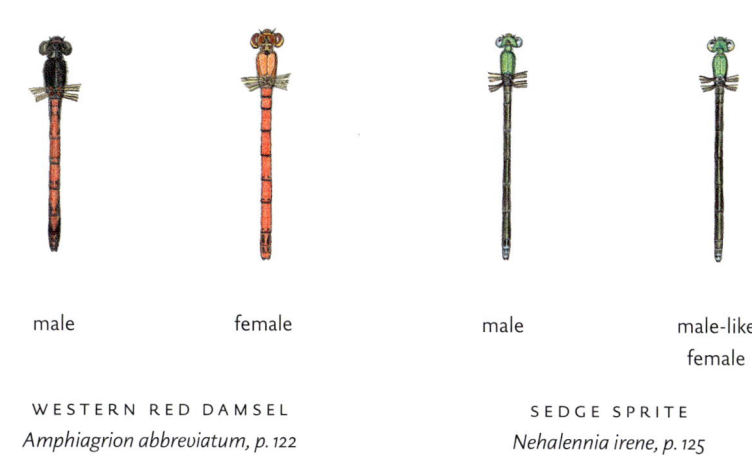

| male | female | male | male-like female |

WESTERN RED DAMSEL
Amphiagrion abbreviatum, p. 122

SEDGE SPRITE
Nehalennia irene, p. 125

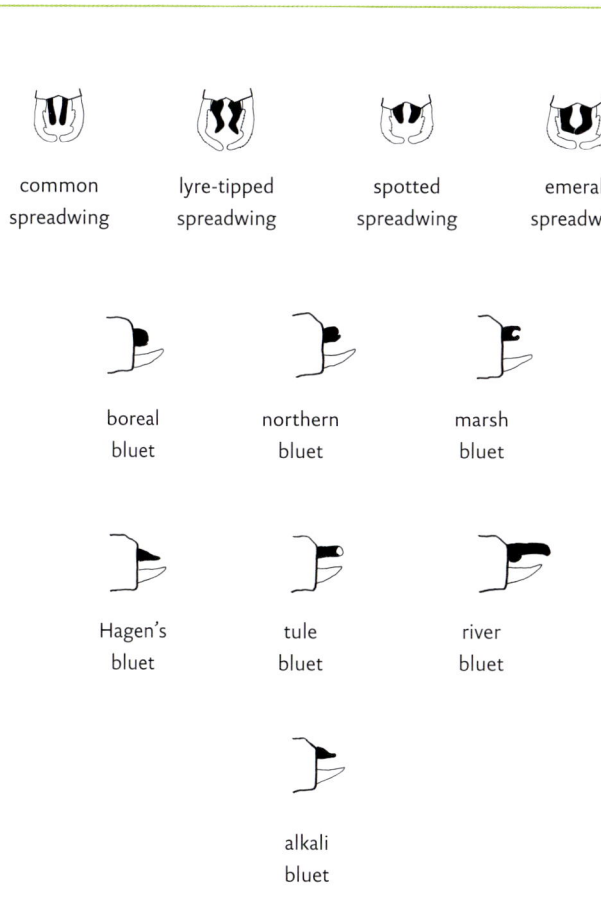

| common spreadwing | lyre-tipped spreadwing | spotted spreadwing | emerald spreadwing |

| boreal bluet | northern bluet | marsh bluet |

| Hagen's bluet | tule bluet | river bluet |

alkali bluet

MALE CLASPERS

boreal bluet	northern bluet	Hagen's bluet	marsh bluet

FEMALE "SHOULDER PADS"

Pacific forktail (tip of abdomen) rear view	plains forktail (tip of abdomen) rear view	western forktail (male claspers) side view	eastern forktail (male claspers) side view

FORKTAIL IDENTIFICATION